GROWING
BOUGAINVILLEAS

Jan Iredell

CASSELL

Acknowledgements

I would like to express again my gratitude to those who have assisted me during my research into bougainvilleas: The Killie Campbell Africana Library, University of Natal, Republic of South Africa; the late Dr. R. E. Holttum; Madame Ohn Sett, Singapore Botanic Gardens; Dr. S. C. Sharma, National Botanical Research Institute, Lucknow, India: Dr. Deepak Ohri, National Botanical Research Institute, Lucknow, India; Dr. Narong Chomchalow, Food and Agriculture Organisation of the United Nations, Bangkok, Thailand; Dr. B. Singh, Indian Agricultural Research Institute, New Delhi, India; Dr. R. N. Bhat, Indian Institute of Horticultural Research, Bangalore, India; Jimmy Dyce, MBE, Loughton, Essex, Britain; the late Peter Greensmith, Nairobi, Kenya, East Africa; the late Mike Lemmer, Harare, Zimbabwe, Africa; Tim Willing, Broome, Western Australia; Philip Wood, Perth, Western Australia; Bill Hatten of Mobile, Alabama, United States; John Lucas of Miami, Florida, United States. Very special thanks to my family for their wholehearted support and encouragement; my husband Peter, for taking all the photographs and without whose help this project would have been impossible; thank you to my son Jon, for making genetics seem so simple; my daughter Penelope for whom the cultivar 'Penelope' is named . . . I thank you all.

Cassell Publishers Limited
Villiers House, 41/47 Strand
London WC2N 5JE

First published in Great Britain 1995
in association with
David Bateman Limited
'Golden Heights'
32/34 View Road
Glenfield, Auckland 10
New Zealand

Distributed in the United States by Sterling Publishing Co. Inc,
387 Park Avenue South, New York, NY 10016, USA

British Library Cataloguing in Publication Data
A catalogue record for this book is available from the British Library

ISBN 0-304-34534-2

Printed in Hong Kong by Colorcraft Ltd

Contents

PREFACE

Most bougainvillea cultivars have more than one name, which confuses both growers and collectors alike. It is not possible to include all cultivars and their synonyms, but listed are over 100 cultivars with general descriptions and photographs of representative varieties. Importations from various countries, all with differing names, are often also given marketing names, which has contributed to this confusion.

Work is continuing on a major book naming and describing cultivars from all countries, with photographs and botanical descriptions, and it is hoped it will not be too long before it is available.

I hope this book will be of interest to all who love bougainvilleas and wish to grow them well.

Note on cultivar names

The Index of Cultivar names (p. 84) provides a quick guide to alternative names for each variety, which in some cases are numerous and can be confusing. These names are listed alphabetically with the synonyms alongside.

Opposite top: Double-bracted cultivar 'Aussie Gold'
Opposite bottom: B. glabra 'Sanderiana'

Hybrid cultivar 'Palekar'

Chapter 1

WHAT IS A BOUGAINVILLEA?

Hybrid cultivar 'Poulton's Special'

BOUGAINVILLEAS are large woody climbers indigenous to the winter-dry tropical regions of South America and they are now grown successfully in most countries of the world.

There are a number of species, but only three of horticultural importance. It is from the interaction of these three species that the modern bougainvilleas have evolved. They flower prolifically and are tolerant of a wide range of soil types and climatic conditions. This adaptability means they can be grown in many forms: from the disciplined bonsai, through potted plants of varying sizes, to small flowering trees (standards), espaliers, clipped hedges, lawn specimens or a tangle of riotous colour when a group of several varieties are planted together.

Bougainvilleas are easy plants to grow as they have no special requirements and are relatively free of pests and diseases. This makes them an extremely desirable and rewarding garden plant.

Bougainvilleas climb by sending out long canes, usually towards the end of the rainy season in their native habitat; the canes are armed with stout, curved thorns. In nature this helps them to penetrate undergrowth and reach full sunlight where they fan out and flower prolifically.

Although armed with thorns, when grown as potted plants the frequent shearing to maintain the desired size and shape keeps new growth soft and thorns to a minimum. This frequent cutting back also promotes constant flushes of new growth and, consequently, flowers, as it is on new growth that flowering occurs.

The brightly coloured 'flowers' that we refer to are in fact bracts, which are leaf-like structures to which the flower is attached to the mid-rib. They come in any number of colours, from white through palest lavender and gold to the deepest of purples, oranges and reds. Many varieties change colour from

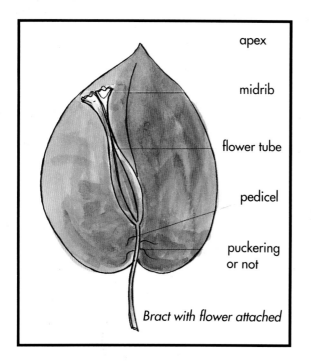

apex

midrib

flower tube

pedicel

puckering
or not

Bract with flower attached

the initial emergence of the bract to its full maturity, often giving the effect of different colours on the one plant. The only colour not represented is a true blue. There are also cultivars which produce bracts of two different colours, for example magenta and white, on the same plant and this has a very striking effect.

There are so many ways of growing bougainvilleas that no matter how small your garden or sunny terrace there will be a bougainvillea which will suit and reward you well with its display of colour for long periods.

Although bougainvilleas are not considered to be especially attractive to birdlife, it is surprising how many birds use them as a source of food, as a safe nesting site and as a protective haven from predators. Australian birds known to frequent the bougainvillea patch include mistletoe birds, honey-eaters and willie wagtails, which nest in them, variegated wrens which frequent them for. a food source, and sunbirds, which feed on the nectar in the flowers.

In Africa, Hunter's sunbirds nest in bougainvilleas and the thickets are frequent-

ed by Superb and Hildebrandt's starlings and finches such as the purple grenadier and cordon bleu; also using the habitat are members of the thrush family and paradise flycatchers. Red cardinals, goldfinches, mockingbirds and larks regularly use bougainvilleas as a habitat in America. In New Zealand, the warblers, fantails, hedge sparrows and silvereyes find the bougainvillea a source of food and shelter. Sunbirds are known to visit bougainvillea flowers in Malaysia, Singapore and India. Hummingbirds are frequent visitors in the bougainvillea's home of South America.

In many places the flowers are also attended by tiny flower spiders which stand poised to pounce on any unsuspecting insect. Nectar-feeding butterflies and moths such as monarchs, white cabbage butterflies, skippers, convolvulus hawkmoths and swallowtails are attracted to the nectaries of bougainvillea flowers. None of these creatures feed on the plant in their larval stages and so cannot be considered pests.

A brief history

This wonderful climbing plant was discovered by the French botanist Philibert Commerson, who saw and described it in Rio de Janiero. Commerson was the botanist aboard the French ship *Bordeuse*, which sailed around the world between 1766 and 1769 under the command of Louis Antoine de Bougainville, the gentleman for whom the genus is named. Bougainville voyaged to discover new territories in the Pacific for France after that country had lost its Canadian empire.

The generic name was initially published as *Buginvillea* in A. L. de Jusseau's *Genera Plantarum* in 1789, 20 years after Commerson made the initial discovery of the plant in Brazil. The name was subsequently spelled a number of different ways until it was finally

Opposite: B. glabra 'Magnifica'

corrected to *Bougainvillea* and published in the *Index Kewensis* (Suppl.9, 1931-35).

Willdenow originally described *Bougainvillea spectabilis* in Linneaus' *Species Plantarum* from the dried specimens collected in Brazil, but there was no detail as to how it grew and there was little to differentiate *B. spectabilis* from *B. glabra*. The original description of *B. glabra* was made by Choisy in 1849, also from dried specimens collected from a number of places in South America. Choisy and Heimerl recognised *B. glabra* as a distinct species and so re-described *B. spectabilis,* which was distinguished from the hairless or nearly hairless *B. glabra* by its hairy leaves, canes, bracts and flower tubes. *B. spectabilis* flowers in response to a cool, dry season, whereas *B. glabra* will flower continuously and profusely in the tropics.

History of cultivation

Bougainvillea spectabilis was the first species introduced to horticulture and was grown and flowered successfully in greenhouses in both Europe and Britain in the early 1800s. It was followed closely by *B. glabra,* which was also successfully grown. At about the same time, the Reasoner brothers of Florida were collecting bougainvilleas from different localities, and these were also *BB. spectabilis* and *glabra*. Other specimens of these species displaying differences in colour and bract size were introduced around this time.

Nurseries in France and Britain specialising in growing tropical plants provided specimens which were subsequently grown in Australia, America and many other places. Kew Gardens distributed many plants to British colonies throughout the tropics and sub-tropics.

The discovery, by Mrs R. V. Butt of Trinidad, of a crimson-bracted bougainvillea in a priest's garden in Cartagena is a major event in the history of bougainvilleas. This plant was thought to be a distinct species

because it had different habits and appearance and it was named *Bougainvillea buttiana* for the lady who found it. However, it was subsequently found to be a natural hybrid and so its correct name is *B.* x *buttiana* (x signifies hybrid origin).

Bougainvillea x *buttiana* proved easy to propagate from cuttings collected in Trinidad, from where it was widely distributed. This lovely plant is still a favourite and widely grown and is nowadays known as 'Mrs Butt'. The original plant of 'Mrs Butt' was introduced into America by the Reasoner brothers in Florida when it flowered amidst a batch of unnamed varieties. It was called 'Crimson Lake' for the pigment colour.

Plants sent to Kew Gardens were raised as stock and were sent to Australia, East Africa, India, New Zealand and other British colonies in 1923.

Professor S. C. Harland later proved that *B.* x *buttiana* was a hybrid when he crossed a variety of *B. glabra* with a 'local pink bougainvillea' in Peru. The latter was almost certainly *B. peruviana,* which did not appear in cultivation until the early 1900s when Mr R. O. Williams reported it as a new pink bougainvillea which had been introduced to Trinidad from Ecuador. It would seem likely that many other hybrid crosses have occurred naturally. During the 1930s when *B. spectabilis, B. glabra* and *B. peruviana* were grown in close proximity, hybrid seedlings began to appear spontaneously in East Africa, India, America, Australia, the Canary Islands and the Philippines, and much interest was shown in these beautiful plants.

Most seed collected was of unknown pollen parentage and many plants now exist which are intermediate between *B. spectabilis, B. glabra* and *B. peruviana*. Chance seedlings grown from seed collected from either *B. spectabilis* or *B. glabra,* or seed collected from a single plant, show a considerable variation in bract shape, size, colour and degree of hairiness on leaves, canes and

B. spectabilis 'Bois-de-Rose'

flower tubes. The seedlings raised from seed collected from *B. peruviana* grow true to type, as does the seed from *B.* 'Alba'.

It was a natural progression that interested growers should begin to experiment and try cross-pollination to see what could be achieved. In 1927 James Hendry in Florida cross-pollinated 'Rosa Catalina' on 'Lateritia', and from this came 'Margaret Bacon' and 'Daniel Bacon', both furry leaved, and 'Susan Hendry', a robust vining type with large crimson bracts, the new growth of which was wine-red in colour and puberulent. Other crosses with 'Crimson Lake' produced 'Barbara Karst', a fine bushy variety with crimson bracts. This cultivar has replaced 'Crimson Lake' in the United States trade. 'Crimson Lake Jr' was one of that group, with compact growth and terminal flowering. Later came 'Helen Johnson', which is still a favourite and widely grown. It is a compact, bushy type

with red bracts, only growing to about 1 m in height and flowering profusely for long periods.

In the 1950s William Poulton in Durban, South Africa decided to plant the three basal species together and started a breeding programme. He produced many hundreds of seedlings, the first of which was a lovely unusual pink named 'Natalii', then came 'Killie Campbell', 'Brilliance', 'Gladys Hepburn' and a dark red with double bracts named 'Wac Campbell'; this was very interesting and quite different in that there were 20-40 small bracts in a cyme and no flowers instead of the usual three. Whether this cultivar and 'Mahara' are one and the same is unproven. All these varieties were propagated by local nurserymen.

Seedlings were also being produced in East Africa by Mr Peter Greensmith and other growers at around the same time. In

Hybrid cultivar 'Daphne McCulloch'

Queensland, Australia, Mr W. F. Turley produced some fine *B.* x *spectoglabra* cultivars.

During the 1970s, at the National Botanical Research Centre in Lucknow, India, work was being done to restore the fertility of selected cultivars by induced tetraploidy. Some notable cultivars from these experiments are 'Chitra' which is a bicolour of rich magenta-red and white, with the added interest of a distinct tinge of gold in the very new bracts, and with a mixture of immature and mature bracts almost becomes a tricolour. 'Mary Palmer Special' was another, and is a softer version of 'Mary Palmer' and much less vigorous in habit. Others with similar characteristics are 'Wadjid Ali Shah' and 'Begum Sikander', both of which have very large, soft bicoloured bracts and a restricted growth habit.

Irradiation of various cultivars has produced even more variations in the form of variegation of leaves, changed bract colour and distortion of leaves and bracts. This procedure has increased the rate of mutation, and consequently there have been up to 50 new bud-sports reported in one year in Thailand.

Natural mutations seem to occur spontaneously throughout the world; wherever large numbers of plants are being produced, bud-sports will occur. This has led to multiple names for the same cultivar and has contributed to the confusion over the names of bougainvillea cultivars.

Chapter 2

THE SPECIES AND HYBRIDS

Hybrid cultivar 'Red Fantasy'

Classification of plant names

PLANT species which show minor differences
from each other are called 'cultivars' or 'varieties'. These differences sometimes occur
between seedlings from the same parent
plants and sometimes as spontaneous mutations or 'bud-sports'. The plant is usually
identified and named by its originator. *The
International Code of Nomenclature for
Plants* covers the special categories and
names of cultivated plants.

The word cultivar is a contraction of the
phrase 'cultivated variety'. A cultivar should
be distinguished from a botanical variety,
which occurs naturally. The cultivar should
have a proper name which propagators can
recognise and which cannot be confused with
the names of other cultivars. Multiple names
for a single cultivar can arise by accident but
sometimes they are created deliberately; in
any event it creates confusion. The most
important principle about cultivar names is

that the first-given name has priority and
should not be changed unless that name is
already taken by another cultivar.

The botanical name of plants includes a
name for the genus, the species and, when
the plant has horticultural varieties, the cultivar. For example :

genus — *Bougainvillea*

species — *spectabilis*

cultivar — 'Thomasi'

This can be written either:
- with the abbreviation cv. preceding the cultivar name — *Bougainvillea spectabilis* cv:
'Thomasi';
- with just quotation marks — *Bougainvillea
spectabilis* 'Thomasi';
- or when the species is not known as
Bougainvillea 'Thomasi'.

If the plant is from a major hybrid group,
then the name may be written as x *buttiana*
(x signifying hybrid origin), for example *B.* x

buttiana 'Mrs Butt'; *B.* x *spectoglabra* 'Sanderiana'.

The species

Bougainvilleas produce massive displays of flowers of unsurpassed brilliance, in an enormous range of colours, in even the toughest conditions. They are plastic in form and can be grown into any shape desired. They are hardy, forgiving and relatively free of pests and disease. The genus *Bougainvillea* has only three species, among a number that are recognised, represented in horticulture. These are *BB. spectabilis, glabra* and *peruviana.* Three major hybrid groups are also recognised, namely *B.* x *buttiana* (*glabra* x *peruviana*), *B.* x *spectoperuviana* (*spectabilis* x *peruviana*), and *B.* x *spectoglabra* (*spectabilis* x *glabra*).

Bougainvilleas in cultivation today come from both natural and fostered hybrids and many more are from spontaneous mutations (or bud-sports — branches with either variegated leaves or different coloured bracts to the parent) among existing types.

Bougainvillea spectabilis

The first species collected, it was described from dried specimens by Willdenow (1798). It is a large, climbing shrub with stout thorns, hairy stems and foliage, and pale corky bark. It has an impressive seasonal display of flowers.

Thorns: woody, strong, hairy and somewhat curved.

Leaves: large, ovate to rounded, somewhat wavy along the margin, leathery in texture and hairy underneath.

Branching: close and short, giving rise to a very dense plant. Inflorescences multiple and terminal in the axillary branches.

Flowering: either during dry season or in response to cool, dry weather.

B. spectabilis

Bracts: large, ovate, obtuse at tip, quilted in appearance and coloured rose, rusty-red and purple.

Flowers: cream in colour, slender, very hairy tubes with fine ribs and a narrow mouth. The tube is only slightly swollen at the base.

B. spectabilis cultivars
'African Sunset'
Not a vigorous cultivar, with furry leaves and thorns of medium size. The bracts change colour from salmon-orange to red, and the flowers are creamy white.

'Alex Butchart'
A vigorous grower with leaves furry to the touch. The rich rosy-red bracts are long and oval, with prominent green veins giving them an almost quilted appearance. Seasonal flowering.

B. spectabilis 'Lateritia'

'Bois-de-Rose'

Beautiful dusty pink with dark green furry leaves and vigorous growth. Seasonal flowering habit. Thorns large and slightly recurved. (A different cultivar is known by this name in the United States.)

'Carnarvon'

Vigorous growth, large furry leaves and large stout thorns characterise this cultivar. Seasonal flowering habit and red colour similar to 'Lateritia'.

'Grimleyi'

The furry oval leaves have long pointed tips on this vigorous cultivar. The medium size bracts are pointed and hang in large drooping bunches giving the plant a shaggy appearance. The bracts are orangey-red and the flowering is seasonal.

'Laidlaw'

The bracts on this cultivar are large and of a particularly seductive colour — the colour of red blush on a peach. It is a vigorous grower with large furry oval leaves, large curved thorns and a seasonal flowering habit.

'Lateritia' syn. 'Dar-es-Salaam'.

An old variety, the parent of many named and unnamed cultivars. Dense, shrubby, slow growing. Bracts medium in size, brick-red colour. Leaves medium sized, furry to touch. Thorns medium, curved. Flowers conspicuous, creamy-white.

'Picta Aurea'

A beautifully variegated form of 'Speciosa'. Variegations in differing sizes and shades, splashed cream and green on the leaves.

B. spectabilis 'Picta Aurea'

'Pink Gem'

The long strong canes on this vigorous cultivar carry large furry oval leaves tapering to a point. The thorns are many, short and slightly curved. The large bracts are bright rosy-pink and have a quilted appearance. The large prominent flowers are creamy white and appear seasonally but for a very long period.

'Rosa Catalina'

Dense, bushy and vigorous. Leaves ovate, furry, tapering to a point. Bracts medium, ovate with pointed apex, uniformly bright rosy pink. Flowers prominent. Seasonal flowering habit. Thorns medium, stout.

'Rubra Variegata' syn. 'Dauphine'.

Compact, rounded, furry leaves tapering to a pointed tip. New growth reddish, marbled with green, pink and cream. Variegation becomes less distinct as leaves age. Bracts large, rosy-red, ovate. Flowers conspicuous, cream. Flowering seasonal; begins early in the season and continues late.

'Speciosa'

Vigorous, dense and rounded. Leaves dull, dark green, ovate, tapering to a short, pointed apex, slightly furry to touch. Thorns medium, stout and curved. Bracts large, rich deep purple, margins slightly wavy. Flowers very conspicuous, cream. Seasonal flowering.

'Splendens'

Rampant variety of *B. spectabilis,* with very hairy leaves, canes and thorns. Leaves large, broadly elliptic to ovate, dark green, tapering to pointed apex. Bracts extremely large, ovate

Opposite: B. spectabilis 'Rosa Catalina'

B. spectabilis 'Thomasi'

with rounded apex, light purple. Thorns long, strong and recurved. Flowers conspicuous and bright creamy yellow. Seasonal flowering habit.

'Thomasi' syn. 'Rosea'.
Dense, vigorous. Leaves ovate, large, medium to dark green, hairy. Thorns large, strong and curved. Bracts very large, carmine-pink, broadly ovate with acute apex, veining prominent. Flowers large and conspicuous, cream.

'Turley's Special'
Dense, vigorous. Leaves medium size, ovate, tapering to pointed apex, velvety to touch. Bracts medium, broadly elliptic with acute apex, cherry-red. Conspicuous cream flowers. Thorns medium to large, curved.

'Wallflower'
Densely shrubby, vigorous. Leaves dark, dull green, ovate, velvety to touch. Thorns medium, curved. Bracts oval, apex rounded, rich bronze-red colour. Flowers conspicuous, creamy white.

Bougainvillea glabra

This is a climbing evergreen shrub with spreading, spiny branches, originally described and named by Choisy (1849).

Thorns: short, thin and curved at the tips.

Leaves: fairly evenly elliptical, widest about the middle, puberulent.

Branching: close with short inflorescences of flowers all along the canes.

Flowering: either continuous or nearly so.

B. spectabilis 'Turley's Special'

Bracts: pointed and triangular, they vary in size. Colours include only white and shades of mauve and purple.

Flowers: conspicuous, tubular, swollen at the base, short hairs (to 1 mm). Corolla white to cream in colour.

B. glabra cultivars

'**Alba**' syn. 'Snow White', 'Key West White'. Pure white bracts with greenish veins, bracts pointed and reflexed. Leaves are long and pointed. Growth in young plants spindly; responds well to regular pruning. Has several flushes of flowers each year.

'Cypheri'
An old variety, rampant growth, leaves dark green and shining, pointed. Bracts large, deep purple; flowers conspicuous and yellowish in colour. Thorns large and curved.

'Easter Parade'
Medium growth habit. Bracts pointed and reflexed, light pink. Leaves ovate with a pointed apex, light green and smooth. Several long flushes of flower in the year.

'Elizabeth Angus'
Vigorous growing; large bright purple bracts; leaves dark green, glossy and tapering. Young stems puberulent; thorns stout, long, recurved. Several long flowering periods per year. Flowers yellowish, large and conspicuous.

'Formosa'
Dark green, pointed, ovate leaves, with sharply pointed apex. Bracts small and pointed, light purple, finally turning to brown, tend to persist on the plant. Flowers conspicuous, tube very inflated and ridged. Very distinctive.

Above: B. glabra 'Alba'

Below: B. glabra 'Formosa'

B. glabra

greenish-cream. Thorns long and curved. Responds well to pruning. Flowers several times a year for long periods.

'Lesley's Selection'
Reported as a bud-sport of 'Magnifica'. The bracts are large, thin-textured and magenta-purple, with dark green, glossy, oval leaves and medium size thorns.

'Magnifica' syn. 'Magnifica Traillii'.
The common purple with dark green glossy leaves. Vigorous, growing into a dense bush. Bracts thin-textured, ovate, brilliant glowing magenta. Flowers prominent, creamy white. Long flowering in summer. Local name for it in Ecuador and Colombia is 'flor de verano' — summer flower.

'Sanderiana'
Loose, upright with small dark green leaves, becoming dense. Thorns many, long and recurved. Bracts ovate, uniformly deep purple, pointed tips, veins green. Flowers conspicuous, creamy-white.

'Show Lady'
Sparse, erect, scandent. Vigorous once established. Leaves light green, broadly elliptic, widest in the middle; petioles long. Bracts hang in drooping panicles, palest lavender-pink. Broadly elliptic with acute apex. Flowers moderately large, cream. Thorns long and slender.

'Singapore Beauty' syn. 'Dr David Barry', 'Singapore Pink'.
Rounded, shrubby, vigorous cultivar, with long elliptic leaves tapering to a pointed tip. Bracts very large, elliptic, lavender-pink and reflexed. Flowers prominent, cream. Thorns short, not prominent.

'Singapore White' syn. 'Mauna Kea White', 'Ms Alice', 'Moonlight'.
White-bracted bud-sport from 'Singapore

'Jane Snook' syn. 'Durban', 'President'.
Leaves long and light green, with long petioles. Compact and dense in habit; tends to be pendulous but does have some vigorous canes. Bracts in heavy bunches, large, pink, ruffled margins. Veining prominently green. Flowers large and greenish cream. Thorns medium and fine. Flowers several times a year for long periods.

'Jennifer Fernie' syn. 'Beryl Lemmer', 'Mudanna'.
Medium grower, somewhat similar to 'Alba' but bracts are broader and leaves narrower. Bracts a lovely pure white.

'John Lattin'
Erect, leggy grower; leaves long, ovate, medium green, glossy and pointed at tips. Bracts typical *B. glabra* shape but tapering to a long point at apex, palest lavender in colour with an iridescent look. Flowers moderate in size,

B. glabra 'Show Lady'

Beauty', identical in form and growth but with white bracts.

Bougainvillea peruviana

This is a climbing, spiny, spreading shrub with greenish bark. It has a similar habit to *B.* x *buttiana,* but it is less robust. *B. peruviana* was described and named by Humboldt and Bonpland (1808).

Spines: thin, straight in youth and curved when older.

Leaves: thin, quite glabrous, ovate to broadly ovate, acute at tip and cuneate at base; petioles long, slender and hairless.

Branching: loose and open; terminal inflorescences.

Opposite: B. glabra 'Singapore Beauty'

B. peruviana

Flowering: recurrent after strong vegetative growth in response to dry weather.

Bracts: small, rounded, delicate and crinkled, pale magenta-pink in colour.

Flowers: tubular and very slender, quite smooth apart from a few hairs on the bracts, yellowish colour. Stamens shorter than the perianth.

B. peruviana cultivars
'Lady Hudson' syn. 'Princess Margaret Rose'. Representative cultivar of this species, long canes often bare of leaves especially when young. Thorns fine and straight. Bracts small, ruffled and pale pink. Leaves broadly ovate with long petioles, light green and glabrous.

B. x buttiana hybrids
These hybrids are large woody climbers with open branches, large dark green leaves and stout, straight thorns. It is similar to *B. peruviana,* but is more vigorous in growth, and exhibits the minimal hairiness (puberulence) of *B. glabra* with bract and flower traits intermediate between both parent species. The

hybrid was described and named by Holttum and Standley (1947).

Thorns: strong, short and straight, rather flattened at the base.

Leaves: very large on main stems, either broadly ovate or slightly heart shaped, apex narrows abruptly to a narrow triangular tip; upper surface dark, dull green with paler veins; copious very short straight hairs on both surfaces.

Branching: loose and open.

Flowering: crowded inflorescences first appear at the end of branches then along canes. Free flowering with flushes of flowers all through year.

Bracts: rounded, cordate base, rounded tip, colours from white through yellow-orange pink to deep crimson.

Flowers: conspicuous in some cultivars, incomplete in others. Complete flowers are tubular, gently swollen in lower half, either cream or tinged with pink. The tube is often much the same colour as the bracts.

B. x buttiana cultivars
'Apple Blossom' syn. 'Audrey Grey', 'Jamaica White'.
Medium grower. Leaves large, rounded, dark green and smooth. Thorns medium. Bracts medium size, white with flush of pink. Very pretty, sometimes shy flowering.

'Golden Glow' syn. 'Millarii', 'Gold Queen', 'Hawaiian Gold'.
Leaves rounded and large. Thorns medium and straight. Bracts broadly ovate, bright gold in colour fading finally to a pinkish-gold. Vigorous and erect in habit. Flowers for long periods several times a year.

'Lady Mary Baring' syn. 'Yellow Glory', 'Hawaiian Yellow'.
Bright yellow bud-sport from 'Golden Glow'.

Growth erect and open, leaves large, rounded, dark green. Bracts medium size, rounded, clear bright yellow. Flowers not conspicuous. Thorns medium, stout and straight. Flowers for long periods several times a year.

'Louis Wathen' syn. 'Orange King'
One of the very early bud-sports from either 'Mrs Butt' or 'Scarlet Queen'. Bracts are rounded, orange in colour changing to bright, fiery rose-pink. Leaves are rounded, large and dark green. Thorns medium, stout and straight. Cultivar is readily distinguished by the incomplete flowers. Tends to be shy-flowering for no apparent reason, but this could be in response to unfavourable conditions.

Left: B. peruviana 'Lady Hudson'

Below: B. x buttiana 'Golden Glow'

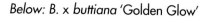

B. x *buttiana*

B. x *buttiana* 'Louis Wathen'

'Mistress Columbine'

Variegated bud-sport from 'Mrs Butt'. The leaves are gold and green with a green central patch. Vigorous once established, but rather slow to start. Bracts and growth habit the same as 'Mrs Butt'.

'Mrs Butt' syn. 'Dame Clara Butt', 'Crimson Lake'.

Habit loose and upright, vigorous. Bracts medium size, broadly ovate and crimson in colour. Leaves large and rounded with a short sharp point at apex. Thorns medium, stout and straight. Flowers prominent, white with a tinge of pink.

'Mrs McClean' syn. 'Hawaiian Orange', 'Orange Glory'.

Bud-sport from 'Mrs Butt'; bright orange bracts of medium size. Leaves large and rounded, thorns medium, stout and straight.

Erect growth habit. Flowers not conspicuous, cream. Distinguished from similar cultivar 'Louis Wathen' by its complete flower tubes.

'Phoenix'

Variegated bud-sport from 'Mrs Butt'. Leaves cream and green, marbled when young but later cream with a green central patch. Slow to start, but moderate to vigorous once established. Bracts and growth habit the same as 'Mrs Butt'.

'Scarlet Queen'

Leaves large, rounded and dark green. Scarlet bracts medium size and rounded. Flowers not conspicuous but are incomplete with little bunches of stamens only. Sometimes in spring there are some complete flowers. Thorns are medium, stout and straight. Growth is erect and open.

'Texas Dawn' syn. 'Purple King'.

Bud-sport from 'Mrs Butt'. Leaves large and

B. x buttiana 'Mrs McClean'

rounded, loose open habit, thorns medium, stout and slightly recurved. Bracts rounded, medium size, light purple-pink. Uniform in colour — no change occurs. Flowers not conspicuous, creamy-white.

The hybrids

Most of the hybrids in cultivation today are of unknown parentage, as they have been grown from naturally pollinated seed. Studies have been done on these hybrids and they have been classified into three broad, distinct groups.

B. x *buttiana* (*glabra* x *peruviana*)

B. x *spectoperuviana* (*spectabilis* x *peruviana*)

B. x *spectoglabra* (*spectabilis* x *glabra*)

The *buttiana* group, represented by the original 'Mrs Butt' was thought at first to be a species as it was so different from *B. spectabilis* and *B. glabra*. It was described and named as *B. buttiana* 'Mrs Butt' by Holttum

and Standley. It later proved to be a hybrid of *B. glabra* and *B. peruviana,* which at the time 'Mrs Butt' was discovered was not known.

The bud-sports from *B.* x *buttiana* are all very similar in form and structure.

B. spectoperuviana has also been studied scientifically and its parentage proven. Representative of this group is 'Mrs H. C. Buck', which has given rise to many bud-sports. These have also proved to be similar in form and structure to the parent plant.

There are other *spectoperuviana* hybrids that have different characteristics to 'Mrs H. C. Buck'. They have varying degrees of furriness and differing flowering habits and bract shape. Without accurate records, one can only make a calculated guess at some of the hybrids' parentage. Knowledge of the parent plants' distinctive features can be a help when trying to ascertain the forebears of a particular cultivar. Many *spectoglabra* cultivars have re-combinations with *B. peruviana* and these show distinctive characteristics.

Spectoperuviana hybrids

'Mrs H. C. Buck' syn. 'Mahatma Ghandi', 'Pacific Beauty'.

Large growing, dense and bushy with long, strong canes. Leaves large, broadly ovate, medium to dark green, glabrous. Young shoots coppery. Thorns large, strong and straight. Bracts large, ovate, margins slightly wavy, deep rich magenta colour. Flowers large, pale cream and conspicuous, appear in recurring flushes.

'Mary Palmer' see **bicolours** for description.

'Shubra' syn. 'Mary Palmer's Enchantment', 'Penelope'.

Bud-sport from 'Mary Palmer' exhibiting white bracts only. Identical in every other aspect to 'Mary Palmer' except for coppery new growth. Foliage lighter green.

'Makris' see **bicolours** for description.

'Thimma' see **bicolours** for description.

'Golden Summers' syn. 'Miss Oneuse'.

Bud-sport from 'Shubra' with green and gold leaves and large white bracts.

Bicoloured cultivars

The first of the bicoloured varieties occurred as a bud-sport from 'Mrs H. C. Buck' after a heavy pruning. This was named 'Mary Palmer' and has subsequently given rise to a number of bicoloured bud-sports. These all have the same base colour but are quite variable in appearance. Later bicoloured varieties were produced from seed of cultivars of the same family that had had their fertility restored. The seedlings raised were part of a controlled breeding programme conducted by the National Botanic Research Institute in Lucknow, India.

'Begum Sikander', 'Chitra', 'Mary Palmer Special', 'Wadjid Ali Shah' are from this programme.

B x buttiana 'Texas Dawn'

'Begum Sikander'

Bracts are large with rosy-purple margins and white centres in the cooler part of the year, but turn to solid rosy-purple in summer. Leaves mid-green, glabrous and elliptic. Thorns slightly curved at tip. Restricted growth. Blooms heavily all along the branches.

'Chitra'

Very strong, upright growth habit, needs regular pruning. Leaves large, dark green, rounded with pointed tips; thick textured. Bracts thick, bicoloured with rich magenta-red and white bracts; emerging bracts have distinct gold colouring.

'Makris' syn. 'Aiskrim'.

Bud-sport from 'Mary Palmer', with the bracts showing pigmentation on the outer parts of the bracts. There are some bracts produced with solid colour, either all white or all pink.

Bicoloured cultivar 'Begum Sikander'

'Mary Palmer'

The first of the bicoloured cultivars. Big grower with bracts of both pink and white on separate branches, intermingled in a panicle or parti-coloured. Leaves large and mid-green, same as for 'Mrs H. C. Buck'.

'Mary Palmer Special'

Softer colouring than 'Mary Palmer', bracts larger and more flaccid. Leaves mid-green, broadly elliptic, thorns moderate, restricted growth.

'Thimma'

Bud-sport from 'Mary Palmer', with the bright bicoloured bracts, but the leaves have a large splash of gold in the centre of each.

'Wajid Ali Shah'

Compact growth, bracts large and magenta-red with irregular patches of white. Leaves broadly elliptic, mid-green, thorns moderate.

Double-bracted cultivars

The first mention of a double-bracted bougainvillea was in South Africa as one of William Poulton's seedlings. Later a complete description was written of the four different-coloured varieties and a single variegated-leaf form. This description was written by Professor Juan Pancho of the Philippines, who also found it as a bud-sport of *B*. 'Mrs Butt', and they were given official names. They have been exported to countries all over the world and have been given many other trade names.

They have no true flowers due to the doubling of the bracts. Each cyme consists of 20-40 small bracts instead of the usual three, resulting in densely packed heads of 'flowers' (bracts). The only disadvantage is that these bracts tend to stay on the plant after fading and can look unsightly. Light pruning will solve this and more shoots will grow and consequently more flowers. There are now five distinct colours. To date, all except the yellow have produced bud-sports with variegated leaves.

'Aussie Gold' syn. 'Carson's Gold'.
The latest colour in the 'doubles' to appear, it arose as a bright yellow bud-sport from plant of 'Roseville's Delight'. It often has red bracts among the gold ones.

'Cherry Blossoms' syn. 'Bridal Bouquet', 'Limberlost Beauty', 'Eastern White', 'Thai Queen', 'Mahara Off-white', 'Tahitian Pink'.
Typical *buttiana* with somewhat erect growth habit and not as vigorous as other cultivars of this species. The 'double' bracts are white tipped with magenta-pink. Thorns medium, with mid green, rounded leaves.

'Coconut Ice' syn. 'Margarita'.
Variegated-leaf form of 'Cherry Blossoms'.
A bud-sport of 'Cherry Blossom'. Bracts 'double', white with pink tips, and the foliage is an attractive cream and green. Moderate grower, slow at first. Medium thorns and rounded leaves.

'Los Banos Beauty' syn. 'Pink Champagne', 'Pagoda Pink', 'Tahitian Maid', 'Mahara Pink'.
A pink bud-sport from the original red double known as 'Mahara'. It also sometimes has odd bracts and odd branches that revert to red.

'Mahara' syn. 'Carmencita', 'Manila Magic Red', 'Klong Fire'.
The original double-bracted cultivar; a deep rich red.

Double-bracted cultivar 'Los Banos Beauty'

A group of five double-bracted cultivars.

'Marietta' syn. 'Cinderella'.
Bud-sport of 'Mahara' with variegated leaves.

'Roseville's Delight' syn. 'Doubloon', 'Golden Doubloon', 'Mahara Orange', 'Thai Gold'. A rich burnt-orange in colour, it occurred as a bud-sport from Mahara.

The other doubles have all had bud-sports with variegated leaves.

Small-growing cultivars

There is much variation amongst the growth habits of bougainvilleas and some of the varieties could be classified as small-growers but not really dwarf. The smallest and most compact that could possibly come into that category is 'Temple Fire' also known as 'Helen Johnson'.

'Dr. H. B. Singh'

Leaves light green and glossy. Bracts large, light, bright lavender. Thorns thin and slightly curved. Compact, rounded habit of growth. Very long flowering period, several times a year.

'Gwyneth Portland'

Leaves ovate and glossy, with short petioles. Bracts medium, broadly ovate, bright yellow colour; several long flushes of flowers during the year. Thorns small and few. New growth coppery.

'Kayata' syn. 'Minyata', 'Shirley'.

Bracts small to medium, rosy-red, twisted in appearance, typical *glabra* shape. Flowers conspicuous and white. Leaves thick, shining, medium to dark green. Thorns small and curved. Flowers almost constantly.

'Little Caroline'

Moderate-sized twisted bracts of fiery orange-red with pointed apices. Leaves small, dark green and pointed. Thorns long, thin and recurved.

Small-growing cultivar 'Gwyneth Portland'

'Mrs Eva'
Moderate-sized, bright mauve, recurved bracts, heavy flowering most of the year. Leaves dark green, tapering to a pointed apex; thorns short, recurved.

'Poultoni' syn. 'Ooh La La'.
Bracts moderately large, ovate with acute apex; opening coppery-red, becoming magenta-red, finally have a bluish appearance. The cream flowers are conspicuous. Leaves broadly ovate, pointed apex, slightly wavy margins. Canes short. New growth coppery-red. Thorns few, small, slightly curved.

'Rosenka'
Bud-sport of 'Poultoni'. Leaves small to medium, somewhat leathery, broadly ovate with pointed apex, light to mid-green. Thorns small and few. Bracts medium-sized, broadly ovate, acute apex, margins slightly wavy; thin-textured, changing from dusky orange on opening to dusky rose-pink at maturity. Cream flowers conspicuous. Recurring flowering habit.

'Temple Fire' syn. 'Helen Johnson'.
The only true dwarf bougainvillea growing to about 1 m tall and wide. Leaves are small, broadly ovate, shortly pointed at apex, margins slightly wavy. Bracts broadly ovate, pointed apex, uniformly reddish-purple in colour, opening slightly coppery. Thorns small, short, slender.

Inter-specific hybrids
'African Cherry Red'
Bracts medium, very twisted, opening a vivid orange-red, maturing to a bright clear red. Flowers conspicuous and creamy-white.

Small-growing cultivar 'Temple Fire'

Leaves small, medium to dark green, broadly ovate with pointed apex, margins slightly wavy. Long flowering period. Thorns thin and curved.

'Alison Davey'

A very showy cultivar, leaves large, flat, mid-green, smooth and pointed. New growth bronze. Habit moderate to large. Has very large flower heads with bracts opening coppery-red and an iridescent magenta. Bracts very large, rounded and ruffled. Flowers prominent and creamy white. Thorns medium and straight. Flowers recurrently.

'Asia'

Medium grower; very good for pots and banks as tends to be rather horizontal in growth habit. Leaves oval with pointed tips and have frequent distortions with silvery-grey markings. Bracts medium size, oval, with distortion in some bracts as in the leaves; brilliant magenta red-purple. Flowers all along the canes, several long flushes of flowers each year. Thorns slender and straight.

'Badgen's Beauty'

Rampant grower with large, glossy leaves and large rich purple bracts in profusion. Thorns

Inter-specific hybrid 'Barbara Karst'

large and recurved. Can be grown in confined spaces but needs firm control.

'Barbara Karst'

Moderate grower. Bracts red and flowers all along the canes; several flushes of flowers during the year. Leaves dull, dark green, broadly ovate with pointed tips, slightly wavy margins. Flowers white and conspicuous.

'Beaujolais'

Rather low growing and spreading in habit. Leaves medium green and rather dull looking. Bracts large, colour of port wine and tend to be seasonal. Excellent for banks or ground cover. Thorns medium and straight.

'Beryl's Red'

Leaves mid to dark green, broadly ovate with smooth wavy margins and a pointed tip. Bracts medium size, thin-textured, slightly coppery-red on opening but overall effect is light red. Habit slightly drooping. Thorns are medium and slightly curved.

'Betty Lavers'

Medium grower, leaves mid green, flat and ovate, pointed at apex. Bracts medium size, ruffled, crisped on margins, tapering to point at the tip; open bright gold, ageing through soft apricot to pale pink. Flowers prominent and cream. Thorns medium and almost straight.

'Black India Orange'

Bud-sport from 'Black India'. Medium growing, leaves very dark green, very rounded with long point at apex, margins crisped. Bracts medium to large, very rounded with pointed tip, margins crisped, brilliant coppery-orange, lighter on opening. Flowers prominent and creamy-white. Thorns medium and slightly curved.

'Black India'

Description same as for above, but the bract colour is bright magenta-red, somewhat coppery on opening.

'Blondie' syn. 'Hugh Evans', 'Fair Lady', 'Delicate'.

Medium grower, leaves light green, narrow and quite distinctive. New growth coppery. Bracts medium, thin-textured, nearly elliptic in shape, pale gold-orange on opening and fading to very soft pink. Flowers prominent and cream. Thorns medium and straight.

'Blushing Beauty'

A bud-sport from 'Juanita Hatten'. Compact growth habit. Leaves two shades of green, giving a slightly variegated look. Bracts light blush pink, ovate, medium size and very prolific. Flowers prominent and creamy-white. Thorns moderate, fine.

'Brilliance'

Medium to strong grower with lax, pendulous habit. Leaves ovate, tapering at each end, dark green with wavy margins; petiole quite long. Bracts ovate with pointed apex and are brilliant fiery-red, new bracts lighter in colour. Flowers prominent and creamy-white. Thorns long and slender.

'Camarillo Fiesta' syn. 'Coral'.

Leaves large, dull, dark green, markedly rounded. Habit strong and open. Bracts very large, rounded and open. Burnt orange fading to coppery iridescent red. Thorns medium and stout. Seasonal flowering habit, more so in dry climates.

'Daphne Mason'

A bud-sport from 'Killie Campbell' with the same large bracts and pendulous growth habit. Bracts lighter in colour, being coppery orange-pink fading to pink. Flowers promi-

Inter-specific hybrid 'Blondie'

Inter-specific hybrid 'Camarillo Fiesta'

nent, large and creamy-white. Thorns large and straight. Leaves long, oval, tapering to long-pointed apex, with a long, slender petiole. Very long flowering in a series of flushes. Moderately vigorous.

'Daphne McCulloch'

Leaves dark green and glossy, margins reflexed. Medium growth habit and tends to be horizontal rather than erect. Bracts round, large and somewhat 'blistered' looking, and are a pure pink. Thorns short and stout. Ideal for ground cover and baskets. Seasonal flowering.

'Donya'

Pink bud-sport from 'James Walker'. Vigorous grower with large, light green leaves with wavy margins. Bracts large and transparent pink, covering the plant all over in recurring flushes of flower. Flowers large, prominent and creamy-white.

'Elizabeth'

Medium growing variety with dark green, ovate leaves pointed at the apex, new growth bronze. Bracts brilliant magenta-red, moderately large; young bracts coppery on opening. Flowers prominent, cream. Thorns moderate and straight. Long flowering periods several times a year.

'Flame'

Loose, upright habit but bushes well with periodic pruning. Leaves broadly ovate, dull dark green; new growth distinctively red in colour. Bracts ovate with acute apex, medium size and good clear red. Flowers cream changing to orange as they age. Thorns medium and straight.

'Gillian Greensmith'

Similar to 'Isobel Greensmith' but lighter colour. Not as fiery.

Inter-specific hybrid 'Donya'

'Gladys Hepburn'

Clear pink, rounded bracts. Vigorous grower, semi-deciduous in spring when it flowers. Flowers prominent, creamy-white. Leaves medium green, dull. Thorns medium and curved.

'Gloucester Royal'

This large-growing cultivar has large, dull, dark green, pointed leaves. The bracts are large and the colour of good claret — the darkest of reds. The large conspicuous flowers are cream. Thorns are large and strong and new growth purplish-red.

'Hensbergii' syn. 'David Lemmer'.

A vigorous, open grower which bushes well with pruning. Leaves large, dull, dark green, rounded. Bracts broadly elliptic, large and bright red. Flowers prominent and creamy-white. Thorns not very prominent. Seasonal flowering.

'Isobel Greensmith'

Loose open growth. Bracts fiery-red, ovate with distinctive pointed appearance. Flowers open cream and become bright orange. Thorns fine, long and curved. This variety tends to be rather spindly when young but bushes well with pruning. Leaves very dark green with pointed apex.

'James Walker' syn. 'Ambience'.

Very large bracts, brilliant red to magenta. Vigorous grower. Flowers prominent, creamy-white. Leaves large, light to mid green with wavy margins. Thorns long and straight. Flowers all over, in recurring flushes. Young bracts very coppery.

'Jayalaxshmi' syn. 'Kiri'.

Dense and bushy. Leaves large, ovate, glossy, dark green, rather thin-textured, tapering to pointed apex; new growth very coppery. Thorns medium to long, slender and straight.

Inter-specific hybrid 'Isobel Greensmith'

Bracts almost round, thin-textured; young bracts shading from reddish-pink at base to red-purple, becoming all-over reddish-purple at maturity. Flowers conspicuous, cream. Pendulous heavy flowering in dry season. Thorns long and straight.

'Juanita Hatten'

Bud-sport from 'Raspberry Ice'. Tends to be ever-blooming with attractive mid green, ovate leaves with slightly wavy margins. Bracts are bright red, ovate and medium to large. Flowers are prominent and creamy-white. Thorns moderate and only slightly recurved. Moderate growth habit.

'Karega Bronze' syn. 'Margery Lloyd.'

Dull, dark green foliage. Drooping in habit but with strong canes. Leaves broadly ovate and slightly furry, sometimes an extra apex on leaves. Thorns medium to straight. Bracts large and thick-textured, broadly ovate, pointed at apex, opening copper, ageing through fiery-red and finally magenta. Several long flowering periods a year.

'Kent's Seedling'

Large growing variety with long, pointed, medium to light leaves, somewhat furry to the touch. The large bracts are quilted in appearance, pointed and light purple. Thorns are

Inter-specific hybrid 'Jayalayshmi'

Inter-specific hybrid 'La Jolla'

long and recurved. Long flushes of flowering several times a year. Flowers large and creamy-white.

'Killie Campbell'

Large growing pendulous variety, young plants have lax habit. Thorns long and straight. Leaves dark green, long, ovate, margins wavy, tapering to long-pointed apex. Petiole long. Bracts large, thin-textured and ruffled, with seductive colour changes from copper through red to magenta. Covers itself all over with masses of flowers in recurring flushes. Flowers large and conspicuous, do not open until late in bract development.

'La Jolla'

Brilliant red, rounded bracts of medium to large size. Flowers prominent and creamy-white. Leaves large, dark green and rounded. Thorns medium and slightly recurved.

'Lavender Queen' syn. 'Lavender Girl', 'Lavender Lady'.

Vigorous cultivar with glossy light green leaves and several long flowering periods throughout the year. Bracts are large and light mauve; sometimes there are extra bracts in the group. Flowers are creamy-white, large and conspicuous. Thorns are large, thin and recurved.

'Mauree Hatten'

Bud sport from 'Barbara Karst'. Growth and descriptions identical to 'Barbara Karst' except for bract colour which is light blush-pink.

'Meriol Fitzpatrick' syn. 'Dream'.

Similar to 'John Lattin' but bushier in habit. Leaves are dark green and elliptic. Thorns medium and recurved. The pale, shimmering lavender bracts have slightly ruffled margins and are thin, ovate and tapering. Flowers are medium in size and creamy-white. Long flowering periods several times a year.

'Miss Manila' syn. 'Tango'.

Vigorous, cascading, dense habit. Leaves

Inter-specific hybrid 'Miss Manila'

Inter-specific hybrid 'Partha'

somewhat ovate, light to medium green; new growth coppery. Thorns medium to strong, straight. Bracts large, whorled, rounded at apex, slightly ruffled, veining prominent. Young bracts open golden-orange becoming a rich, soft apricot colour and fading to light pink, giving a multi-coloured look. Recurrent flowering habit. Flowers large and prominent, cream.

'Mona Lisa'
Moderate growth habit. Very dark green leaves, rounded, tapering to long point, edges crisped and reflexed; leaves look as if the edges have been rolled. Bracts are rich magenta-red, large, rounded with crisped, rolled edges. Flowers are prominent and creamy-white. Thorns moderate and slightly recurved.

'Natalii'
Drooping branches and fragile growth habit. Bracts large and twisted in arrangement, soft dawn-pink with midrib prominently green;

bracts borne in huge pendulous clusters. Leaves long and light green. Thorns small and sparse. Dried bracts persistent.

'Nina Mitton'
Bushy, dense, moderate growth. Mid green leaves are thick, shining and ovate, with a short petiole. Bracts are small, almost round, with a sharply pointed apex. Their soft rose-pink does not change with age. Seasonal flowering habit, covers itself with colour.

'Palekar'
Low-growing variety suitable for banks or ground cover. Leaves small to medium size, dark green, ovate, tapering to a pointed apex, petiole long. Bracts small to medium, dark red, ovate with pointed apex. Flowers not very prominent. Thorns thin and medium.

'Partha' syn. 'African Star', 'Indian Flame', 'Lemmer's Special'.
Tends to sprawl, which makes it an ideal subject for banks and covering large areas. Bracts

thick, long, pointed and slightly twisted; flame coloured, opening coppery-orange, through red to magenta. Leaves ovate, tapered at both ends, light to medium green and slightly rough to touch. Thorns medium, straight and strong. Flowers prominent, large, creamy-white. Recurring flushes of flower.

'Pink Clusters'

Bushy variety with small, rounded, dull, dark green leaves pointed at the apex. Bracts are small to medium, prolific, slightly twisted and with pointed apex; soft dusky-pink colour, fading to pinkish-beige at maturity. Flowers smallish, cream. Thorns inconspicuous, short and slightly recurved.

'Pink Pixie' syn. 'Hawaiian Torch', 'Smarti-pants'.

Medium to tall, upright plant with small leathery leaves, internodes very short and leaves crowded. Thorns short, stubby and blunt, larger on strong canes. Bracts small and densely packed, magenta-red in colour. Flowers prominent, small and white.

'Poulton's Special' syn. 'Starfire', 'William Poulton', 'Clifton Carnival'.

Moderate grower, sturdy and compact in habit. Leaves broadly ovate, thick, glossy, dark green with wavy margins; new growth

Inter-specific hybrid 'Pink Clusters'

Inter-specific hybrid 'Pink Pixie'

reddish and slightly puberulent. Bracts large, broadly ovate, twisted and ruffled, rich rosy-magenta. Flowers large, cream, distinctly ruffled. Thorns moderate.

'Pride of Zimbabwe'

Not vigorous but bushy and tends to be deciduous. Bracts are medium in size and light crimson, maturing to a rosy claret colour. Leaves slightly furry to the touch and mid-green. Thorns medium and slightly curved. Conspicuous cream flowers.

'Purple Robe'

Large, strong, vigorous climber. Dense and bushy with puberulent canes. Dark green leaves are glossy, ovate and pointed at tip, puberulent underneath. Bracts are large, ovate, pointed and puckered at attachment to petiole. They are brilliant purple and sometimes extra bracts and flowers give a tasselled look.

Inter-specific hybrid 'Sakura'

'Rainbow Gold'
Bud-sport from 'Mrs McClean'. Bracts a softer shade of orange, otherwise characteristics same as 'Mrs McClean'.

'Red Fantasy'
Bud-sport from 'Juanita Hatten' with gold-splashed leaves. Bracts and growth habit same as 'Juanita Hatten'.

'Sakura' syn. 'Flamingo Pink'.
Moderate grower with dark green leaves, broadly elliptic and with long petioles. Bracts medium, ovate, twisted in arrangement, white with distinctive shading of pink on upper half of bract. Flowers small, white. Thorns moderate. Recurrent flowering habit.

'Scarlet Glory'
Moderate grower with distinctive reddish new growth. Leaves dark green, broadly ovate,

tapering at both ends, petiole short. Bracts medium, ovate, light rosy-red. Flowers small, conspicuous. Recurrent, profuse flowering habit. Thorns moderate and recurved.

'Scarlet O'Hara' syn. 'San Diego Red', 'Hawaiian Scarlet'.
Loose, upright, vigorous, with strong canes. New growth dark red in colour and slightly puberulent. Leaves very large, rounded, dark green, thick-textured. Bracts nearly round and very large. Young bracts slightly orange shaded dark red. Flowers conspicuous, cream, large. Thorns many, medium, recurved. Seasonal flowering habit, often on bare stems in the spring.

'Srinivasia'
Vigorous, cascading, dense habit. Leaves very large, mid to dark green, smooth, ovate tapering sharply to a pointed apex; new growth

42

bronze. Thorns large, stout and slightly recurved. Bracts very large, ovate with acute apex, margins wavy, brilliant rose-pink to red. Flowers conspicuous, cream. Recurrent flowering habit.

'Tetra Mrs McClean'

Tetraploid version of 'Mrs McClean'. Moderate growth habit. Leaves rounded, dull, dark green, sharply pointed and thick-textured. Bracts more richly coloured than original 'Mrs McClean', thicker textured. Thorns medium. Sets seed.

'Treasure'

Erect and shrubby habit, with medium, stiff growth. Leaves medium, thick, waxy in appearance, mid green, ovate, tapering to pointed apex. Bracts medium, densely crowded, pointed, opening a coppery-pink, ageing to light magenta-red. Flowers conspicuous, white.

'Vera Blakeman'

Shrubby and erect habit with stiff growth, medium grower. Leaves medium, thick, waxy in appearance, mid green, ovate tapering to pointed apex. Bracts medium, densely packed, pointed, uniformly magenta-red. Flowers conspicuous, white.

Variegated cultivars

Variegated bud-sports have occurred in almost all bougainvillea cultivars. They arise naturally wherever large numbers of plants are propagated. In recent years irradiation of various varieties have produced large numbers of variegated bud-sports. Some of these have been individually named, others have not. There is much variation in the nature of the variegation, ranging from a few spots and streaks to large patches of cream or gold in the centre of the leaf blade. The majority are marbled in several shades of green and cream or gold. They are an exciting development, adding much variation to bougainvilleas.

Inter-specific hybrid 'Scarlet O'Hara'

A tetraploid of *buttiana* 'Mrs McClean'

'Bhaba' a bud-sport from 'Louis Wathen'
'Blondie' variegated from 'Blondie'
'Brilliance Variegata'
'Double Delight' a bud-sport from 'Barbara Karst'
'Golden Summers' syn. 'Miss Oneuse' from 'Shubra'
'Harrissi' from *B. glabra*
'Jamaica Orange' syn. 'Freckleface' from 'Jamaica Red'
'Jamaica Red' from 'Mrs Butt'
'Jamaica Yellow' from 'Jamaica Orange'
'Marietta' syn. 'Cinderella' from 'Mahara'
'Miss Manila' variegated from 'Miss Manila'

'Ninja Turtle' from 'Pink Pixie'
'Orange Stripe' from 'Lady Mary Baring'
'Ratana' syn. 'Butterfly': red/mauve/orange/pink/yellow/white' Irradiated cultivars with attractively distorted bracts which look like butterflies. Leaves variegated and distorted.
'Red Fantasy' from 'Juanita Hatten'
'Scarlet Pimpernel' from 'Scarlet O'Hara'
'Scarlet Queen' variegated from 'Scarlet Queen'
'Thimma' from 'Mary Palmer'

There are many others too numerous to list in this publication.

Inter-specific hybrid 'Vera Blakeman'

Environmental colour changes

The bracts can change colour in different environmental conditions, which often makes the identification of cultivars confusing. Plants grown in full sunshine and low humidity have bracts with more intense pigmentation than plants grown under humid, tropical conditions in a glasshouse. Colder weather also causes more intense colours.

The magenta colouring of the bracts consists of bands of red-purple pigments (betacyanins) and yellow-orange pigments (betaxanthins). Light intensity influences the deposition of these pigments, betacyanins being the most responsive. Thus a rich magenta-red bougainvillea grown in full, clear sunlight will be a much richer and brighter colour than the same variety grown in a glasshouse or humid, tropical climate.

The glasshouse plants could be quite pale pink, whereas bright purple plants grown in sunshine will be pale mauve in a glasshouse.

The light is more diffuse in humid, hot climates than in dry air. Plants grown indoors are paler, as are those grown in the shade, because they get less light and deposit less pigment in their bracts. It is interesting that the white, gold, orange and red bracts of *B.* x *buttiana* cultivars all show a tendency to become purple as they fade. This appears as tinges of purple at the tips and margins of the bracts, particularly those exposed to the maximum amount of light. *B. spectabilis* has bracts which change to a more coppery tone as they fade. The bracts on *B. glabra* varieties change little with age but can produce more red tonings when they fade.

Chapter 3

GROWING BOUGAINVILLEAS

Variegated cultivar 'Bhaba'

WHEREVER you might like to grow a bougainvillea, there will be a plant suited to your requirements, and because of their adaptability even large-growing varieties can be successfully grown in containers. Bougainvilleas are very plastic and can be trained to grow in a variety of ways. They can be grown in pots and baskets, as shrubs and bushes, through trees, or to cover a wall or bank. Many of the modern cultivars have differing growth habits and sizes; this means less effort is involved in attaining and maintaining the desired effect if you choose your cultivars carefully. In the following pages you will find detailed information on how to raise and train your bougainvilleas.

Selection of the individual plant is important. Small plants are generally more fragile than larger ones, which are also a little more mature. The plant should be sturdy and have thick, white roots which should be emerging from the drainage holes or should at least be visible. If the plant you are buying is coming straight from glasshouse conditions, it is advisable to provide it with some shelter until it is fully hardened off (see page 56); it is not advisable to plant it straight into the garden.

Bougainvilleas do best in a sandy loam soil — one that does not become waterlogged. For quick growth to become established, plentiful water and fertiliser should be given. Once established, both should be reduced to control the rate of growth. This does not mean that the plant should not be fed and watered at all, however.

Soil

Soil science is a complex and specialised subject, but you need to know something about it because the well-being of your plants depends on the quality of the soil in which they grow.

Soils are classified as varieties of clays, loams and sands. Loam is the best medium to grow plants in because it combines clay, silt and sand together with humus to make friable (crumbly-textured) soil. The most important factor in creating friable soil is the presence of humus, which is produced from organic matter by the actions of micro-organisms and invertebrates living in the soil. Soil organisms are vital to help form well-structured soil; algae, bacteria, fungi, insects and worms all work to build and improve your soil. Humus in soil is more advantageous to plants than its small proportionate amount would indicate. A sound rule of thumb in gardening is to be generous with organic manures and careful with inorganic fertilisers.

Plants grow best in crumbly soils like loam because nutrients and oxygen are freely available to their roots. The greater the depth of soil, the deeper the roots will be able to penetrate and the greater the amount of nutrients and moisture available. The result is a vigorous plant which can withstand periods of drought.

Water

Water is essential to plant growth, most of it being absorbed through the root hairs, although a little is absorbed through the leaves. The importance of water may be appreciated when it is realised that 70-90 percent of the plant's weight is water. Water is not only used in the manufacture of plant foods, it is also used in transpiration and when this exceeds water absorption, wilting results. Excessive dryness causes root death, which obviously retards plant growth.

Bougainvilleas are broad-leaved plants and being vigorous growers need plentiful water when making new growth. For a good display of flowers it is essential that the plant makes continual new growth — no new growth, no flowers. Unfortunately there is no hard and fast rule regarding how often you should water, such as every second day or once a week. However, plants should only be watered when the soil surface has become

dry. The amount of water needed depends on the soil type, climatic conditions, size of the plant, whether it is growing in a pot or in the ground and the amount of humus and other moisture-retaining matter surrounding the plant.

Deep but less frequent watering is far better for your plants' good health than frequent light sprinklings. Surface watering encourages the development of surface roots, whereas deep watering will encourage roots to penetrate deeply into the soil and so firmly anchor the plant. Hot, dry, windy weather requires more water to be given as the rate of transpiration is increased (i.e., water lost from the surface of the plant's bracts, leaves, etc.).

Plants grown in containers need their water requirements attended to more closely as it is very easy for them to use up all the available moisture in their pot. In rainy weather, unless it is particularly long and heavy, the plants will not get enough water to satisfy their needs and so will still need watering.

B. glabra 'Singapore White'

Light

Bougainvilleas are sun-lovers and hence require as much sunshine as possible for optimum growth and flower display. Low light and shady positions are not at all suitable for bougainvilleas; they will drop their bracts and leaves in protest. They must have bright light at the very least.

They flower best as outdoor plants but they can also be grown successfully in conservatories and glasshouses; glasshouses mimic the tropical climate of its native range. Very bright sunrooms or window alcoves which receive the sun for most of the day are suitable. The bracts will be a lighter colour than those grown in full sunshine, but are still bright and beautiful.

Position

The position and preparation of the site selected for planting are most important for getting the best results from your bougainvillea. Choose an open, sunny position, preferably sheltered from cold winds; bougainvilleas which grow in full sun are healthy and will flower prolifically. They will survive in shade but only flower sparsely if at all, and develop a spindly and leggy growth habit.

You should think about how your young plant will look when it is mature and allow enough space for it to develop properly. Bougainvilleas are long-lived plants and some varieties can grow to a prodigious size given optimum conditions. If planted too closely with other plants it may grow up over the top of neighbouring plants, spreading out and flowering there, but in so doing may crowd out its neighbours. Regular pruning will control the plant's size and shape. The majority of active growth occurs from the middle to the end of summer, when many strong watershoots are produced; this is the time to work on your bushes to bring them back to shape by removing watershoots and cutting back canes that have grown too long (see Pruning, page 54).

Selection of varieties

Bougainvilleas are considered hardy plants and this is indeed so, especially once they are established. However, some varieties are hardier than others. *B. glabra* and *B. spectabilis* varieties are most tolerant of temperature extremes, and will even tolerate light frosts. On the other hand, *B. peruviana* is definitely a tropical plant and consequently some hybrids are more suited to warmer climates, unless grown in a conservatory or glasshouse. Here, of course, they can be grown to perfection. If you are growing bougainvilleas in cold climates, these varieties are most likely to defoliate during the winter. Although they require little water during this time, it is important to keep the soil moist (not wet) — don't let it dry out for any extended period.

The selection of varieties for size should also be considered. The variegated cultivars are less vigorous than the green-leafed varieties and are also slower growing. This means that they are more suitable for smaller areas and containers. A list of smaller-growing varieties and their characteristics are listed on pages 31-32.

The point to remember about most hybrids is that many of them have inherited the best characteristics from their parent plants, i.e., they flower prolifically and often in an enormous range of colours, so there is a bougainvillea for any situation and colour scheme.

Site preparation and planting

Bougainvilleas tolerate most soils but must be planted out carefully. Good preparation gives them a good start in life, and this helps them to become strong and healthy plants. This in itself means less fuss with the plant later. The hole should be at least half as deep again and twice as wide as the container that the plant is already in. Fill the hole with water to assess drainage and if it acts like a reservoir, break

Opposite: B. spectabilis 'Speciosa'

away the sides to promote drainage. In heavy soil, digging the hole with a spade makes the sides of the hole very smooth and they act like a dam. By breaking away these sides, you allow the water to escape into the surrounding soil. Bougainvilleas hate to have wet feet and their roots will die if they are left to sit in a waterhole. Fill around the plant with a mixture of good sandy loam, animal manure, compost and slow-release fertilisers, either in pellets or granules.

When you are planting out either a long way from hoses and taps or where your plants will get only infrequent watering, you can use moisture-retaining agents which are available from most nurseries or garden supply centres. Mix the reagent to a gel and place in the bottom of the hole under the root-ball of the plant. These agents can hold up to and above 4000 times their volume of water; this can be a saviour in times of drought. Moisture retainers will last through the growing season, which should be enough time for the plant to establish a good root system. This does not mean, however, that you never have to water the plant — it means that the plant is simply more likely to survive less frequent watering in the early stages after planting out.

If you keep your plants in pots for several weeks before you plant them out, remember that they have been well nurtured in the nursery before you bought them, so be sure to check them daily and water well when necessary. Keep them in a sunny position until you plant them out. When you do plant them out, make sure that the root-ball is moist. Allow all potted plants to drain thoroughly after they are watered and before you begin to plant or to re-pot them, otherwise the weight of the water in the soil could cause it to collapse when the plant is transferred, and so damage the roots.

Do not plant too deeply; the plant should be at about the same level in the ground as it was in the container. The soil around the newly planted bougainvillea may be firmed by gentle pressure but do not stomp it down with hobnail boots; tread gently and softly. Apply a suitable mulch to the whole area around the new plant and water thoroughly, preferably with a sprinkler.

Spacing

Give your plants enough room to grow to maturity without being crowded. Most varieties are comfortable at between 2-3 m apart. Smaller varieties can be planted closer together.

Staking

Bougainvilleas are large growers and can develop bulky crowns, so it is advisable to tie them to a solid hardwood stake. The stake should be driven into the ground before planting as putting the stake in after planting can damage the roots. Use short stakes when you want cascading growth, as in window boxes and down banks. The heads of bougainvilleas can become quite large and such bushes can be snapped off at ground level in very strong winds.

Transplanting and aftercare

Select and prepare your new site carefully in advance; do not transplant bougainvilleas in either autumn or winter. Spring is the best time, after the winter chills have gone and before it becomes too hot. Remember not to skimp on site preparation otherwise you may forever be trying to put it right. Plants frequently fail after they are transplanted, either because their roots have been damaged or the gardener failed to provide adequate care to help the plant to become re-established. Prune the top of mature plants back by about one-third to one-half and cut in a circle around the plant with a sharp spade at about the circumference of the new dripline (area on the ground matching the perimeter of a plant's leafy canopy), or about 30 cm for each 2.5 cm of trunk thickness. Keep the spade and the cut vertical, not at an angle. This cut

Inter-specific hybrid 'Alison Davey'

should be made several weeks before the plant is moved and the plant should have been fed with an agent containing root-inducing hormones to stimulate feeder root growth.

Take great care when the plant is finally in place. Stay wires will help to hold large plants; they should not wobble while they are establishing a root system. Water every other day for about 5-6 weeks and spray the foliage frequently to further reduce water loss.

Mulch

Mulch helps to retain moisture in the soil by reducing evaporation. Spread it 10-15 cm deep around the plant but keep it at least 10 cm away from the trunk or stem to avoid the possibility of fungal attack.

Material that can be used as a mulch includes wood chips and bark (though be sure to check that the wood it came from wasn't treated with chemicals), leaf litter, compost, stable straw, spent mushroom compost, chicken litter (very good, as long as it is sufficiently decomposed) or well-washed seaweed. A sprinkling of nitrogenous fertiliser (i.e., urea or sulphate of ammonia) over the mulch if it contains decaying vegetable matter is advisable. Otherwise the mulch will use up the nitrogen in the soil as it decomposes, as it is an essential element in this process, meaning your plants won't get a look in.

Lawn clippings do not make a good mulch, unless they have been combined with other organic matter and composted. When used straight from the lawn mower, or after being heaped and dried, the clippings form a mat that acts as a barrier preventing any water penetrating to the soil below.

Nutrition

Plants, like all living things, need food and water to survive. Bougainvilleas, although hardy plants, are no exception. Plants produce much of their food by way of photosynthesis from relatively simple compounds like water and carbon dioxide which they combine in the presence of sunlight to make carbohydrates. Water and salts (minerals in solution) are absorbed from the soil to provide the requirements for building plant structures. Fertilisers contain many of the essential nutrients in organic and mineral form. Bougainvilleas need fertilising regularly – every three weeks – with an evenly balanced fertiliser (NPK 8:8:8 is ideal, i.e., nitrogen, phosphorus and potassium are in equal proportion).

Plants absorb the substances they need in soluble form through their roots. Plants starve when there is a shortage of any one nutrient and individual deficiencies show up as different signs and symptoms. Try to learn the subtleties of these signals from your plants before jumping to conclusions. In addition, remember that the elements missing in the plant may not necessarily be lacking in the soil; it may be that they are not in a form which is available to the plant.

The availability of nutrients is closely tied to soil pH. This is a measure of acidity on a scale ranging from 1-14 where 1 indicates extremely acid conditions, 7 is neutral and 14 is extremely alkaline. Soils range in pH from about 3 in peat bogs to 10.5 in arid, alkaline soils. Most plants grow best at pH 6-7 and at these pH values all the minerals which are essential to plant growth are soluble and available.

Bougainvilleas tolerate a wider pH range than many plants, but they may show iron and magnesium deficiencies (chlorosis), especially in pots, just after flowering, at the end of winter and in the tropics after prolonged rain.

Common inorganic deficiencies include:

Nitrogen (N) Older leaves turn a yellowish-green with veins sometimes reddish. Severe deficiency shows as stunted young leaves and slender, hard young twigs.

Phosphorus (P) Veins and leaf stalks turn red to purple and the leaf blade becomes mottled yellow to blotchy brown.

Potassium (K) Edges on older leaves become purplish, generally preceded by tipburn; eventually the younger leaves are affected.

Magnesium (Mg) Older leaf blades become either blotched yellow or tan with veins remaining green.

Zinc (Zn) Can be similar to magnesium but leaves may be stunted, crinkled and their growth crooked.

Sulphur (S) Younger leaves turn entirely yellow, with the veins a brighter colour than the leaf blade.

Molybdenum (Mo) Mottling over whole leaf but little pigmentation; leaves cupped with distorted stems.

Manganese (Mn) All younger leaves remain green, blades turn yellow and later brown blotches appear over entire leaf.

Calcium (Ca) Dead areas appear in young tender leaves at tips and margins. Terminal buds remain either smaller than usual or shrivel and die. Calcium becomes unavailable if the soil pH is greater than 7.8.

Boron (B) Terminal buds appear similar to those suffering calcium deficiency; leaf stems brittle and leaf bases turn yellow.

Iron (Fe) and Aluminium (Al) Often occur when soil is too alkaline; leaf blades either

'Jane Snook', 'Scarlet Glory' and 'Tango' look spectacular together.

pale or yellow, larger veins remain green but smaller veins turn yellow; later, edges of leaves turn brown.

When the major elements (N, P and K) are lacking, an application of a standard chemical fertiliser should overcome the problem. If minor elements are missing, specially mixed formulas of soluble trace elements may be used. Single element additives, such as iron chelates, are also available.

Remember that too little is better than too much. Many healthy young plants have been over-fertilised and killed. Natural products, such as fertilisers made from seaweed, fish products, compost and animal manures, can supply all elements needed for plant growth. Either a soil test or leaf analysis can be done to establish the proper treatment needed.

Chlorosis

Chlorosis is the term used to describe the yellowing of leaves commonly caused by mineral deficiency, most commonly of iron, magnesium and nitrogen. Most bougainvilleas with chlorosis respond favourably to watering with iron chelates together with magnesium at the recommended rate at 10-14 day intervals until the leaves become green.

Manures

Comparative Values of Animal and Poultry Manures			
Source	Nitrogen (%)	Phosphoric Acid (%)	Potash (%)
cattle	0.20	0.17	0.10
horse	0.44	0.17	0.35
sheep	0.55	0.31	0.15
pig	0.60	0.41	0.13
poultry	1.10	0.85	0.56

Poultry manure is richest in elements but lacks bulk and humus. Manure makes wonderful mulch which is full of essential nutrients; fresh manure is best but always keep it away from the base of plants. 'Weathered' manure has had much of the nutrients leached out of it.

Many nutrient deficient plants do not grow well and have poorly developed root systems; it is sometimes possible to save these plants by potting them back to a smaller pot but often it is better to discard this specimen and start again.

When buying your bougainvillea check the drainage holes to see that there are good strong roots showing through. Also the plant should be firm and not wobbly in the pot.

Foliar fertiliser

Foliar fertilisers are either poured or sprayed over the plant in a solution so that nutrients can be absorbed through the leaves. A small amount (1 teaspoon to 10 litres of water) of household detergent or white oil helps the fertiliser 'stick' on the leaves and promotes absorption. Foliar fertilisers can be used in association with other fertilisers and can be applied every 2-4 weeks in the growing season. Many commercial liquid fertilisers are available and any of them are suitable.

Iron chelates, potassium nitrate, magnesium sulphate (1 teaspoon to 4 litres of water) could be included in alternate feedings. This will help to prevent the yellowing of leaves known as chlorosis.

Pruning

There are no hard and fast rules for the pruning of bougainvilleas, but for the best results, pruning should be carried out after flowering has finished. This encourages new growth on which the next flush of flowers occurs.

If you want to reduce the size of the plant, cut back by about a third, removing all spindly and twiggy growth. Being good climbers, pruning bougainvilleas regularly is important if you want to keep their size under control, particularly for container plants and those grown in conservatories.

Watershoots, which arise from woody stems after hot, wet weather, are strong, fast growing, sappy shoots and are the means by which bougainvilleas grow up through the foliage and reach the canopy, and the light, in their natural habitats. These watershoots should be cut off close to the stems if they are not wanted. However, they can be used to advantage if you are creating a 'standard' bougainvillea or growing a plant over a pergola — in fact anywhere a straight stem and height are required.

Note that the varieties of *B. glabra* and *B. spectabilis* will respond better to hard pruning than will cultivars of *B.* x *buttiana*. A good general rule is that regular light pruning of your bougainvilleas will keep them in good shape with plenty of new growth and flowers.

Holiday care

All gardeners face the common problem of how to care for their plants when they go away on holiday. If a friend or neighbour cannot help there are a number of ways to solve this problem.

An automatic watering system is a good investment and is the way to ensure proper plant maintenance. They also save a lot of time, and the advent of small, computerised, home watering units is wonderful! They can be programmed to deliver the exact amount of water required for your particular garden. Timer taps are less expensive and quite useful, as they will automatically turn off after the preselected time has elapsed.

There is a product available called 'tree-bags'; these are large plastic bags which hold about 10 litres of water and have an adjustable wick outlet to control the volume released. The bag is placed at the base of the plant and the water will trickle out for up to three weeks. The bags need to be well covered with mulch or the water released will be hot and can damage the roots. Do not use these bags as permanent fixtures as they can waterlog the ground or cause the roots to be directed only to a small area where the soil is moist. The old adage of flooding regularly and allowing to dry — but not completely — is still the best to keep your plants healthy; this keeps the roots expanding outwards and deeper, seeking water and nutrients.

Cultivation in cool climates

The secret of successful gardening is to grow plants in an environment that is as similar to their natural habitat as can possibly be attained. Bougainvilleas are hot climate plants, with their natural habitat in tropical and sub-tropical South America. They suffer in the cold but it is still possible to grow them and enjoy the beauty of their flowers in such climates. Being very adaptable, bougainvilleas can be grown from cool climates, i.e., northern Europe, the northern parts of the United Kingdom and the United States, through temperate regions in the south of these countries, Australia, South Africa and New Zealand, to tropical habitats close to the equator. If grown outside, they grow best in areas that have free-draining soils, hot, dry summers and frost-free winters. In cool cli-

'Poultoni' is a popular small-growing cultivar.

mates, they are very successful indoor plants if placed in bright, sunny rooms, and they make excellent conservatory specimens. When grown indoors, bougainvilleas are often trained as small, flowering pot plants or as basket plants. If you wish to grow larger plants indoors ensure that you plant them in suitably large pots.

Outdoors in frosty areas bougainvilleas will become deciduous, defoliating completely during winter. As long as the days are warm and sunny, this causes few problems, especially if the hardier varieties are chosen. However, prolonged heavy frosts and snow precludes the growing of bougainvilleas outdoors. Many gardeners overcome this by growing their bougainvilleas in large pots (good plants can be grown in 20-cm to 30-cm pots) and moving them around to the warmest position in the garden, or bringing them inside for the winter. If you intend to move your plants, remember that plastic pots are lighter and easier to handle than clay ones, and if they are dark in colour they absorb more heat into the soil during sunny winter days. On the other hand, this does mean that in summer it is best to put the plastic pot inside a clay pot as a 'sleeve' to prevent the roots from getting too hot.

The potting medium should be open, free draining and rich in humus and do not let the pots stand in saucers full of water in the winter months.

Don't put your plant beneath a tree for protection from frost, because the shady conditions will be equally harmful. Where night temperatures consistently drop below 4°C (40°F) bougainvilleas are best wintered in a glasshouse or, alternatively, in a sunny glassed porch or a sunny room. Keep them there until warm weather returns. If the winters are cold enough for snow, a heated glasshouse will be required. Glasshouses can be freestanding or attached to a house. They can be made of plastic, fibreglass or glass. The glasshouse must receive the maximum amount of sunshine and it should be positioned accordingly.

When moving plants outside from a glasshouse it is necessary to gradually harden off the plant, otherwise the leaves and flowers could be scorched. The reason for this is that the cell walls in the leaves are thinner in plants kept in glasshouses so that they can absorb the maximum sunlight for photosynthesis. (Photosynthesis is the process by which plants and parts of plants with chlorophyll (the green parts of plants) use light energy to convert basic inorganic elements such as nitrogen and carbon dioxide into sugars, or 'foods', that the plant can use to grow and reproduce. This fundamental process converts inorganic materials into organic materials that living things can use.) Plants grown in glasshouses are often called 'soft', compared with 'hard' plants growing in the open. Keep glasshouse plants for several weeks in a 'halfway house' before you move them outside; this will give the cell walls and cuticle time to thicken and strengthen. The light, dappled shade of a tree sheltered from strong winds is excellent for this.

Frost protection

Bougainvilleas need to be protected from frost when they are young. The best protection is to cover them up at night to trap daytime warmth and ground heat. Bubble-plastic sheeting makes the best blanket as it lets more light in, but an ordinary clear plastic sheet will do. Make the plastic shelter large enough so that the leaves do not touch the plastic itself and risk of sunburn is reduced.

Chapter 4

USING BOUGAINVILLEAS

B. x buttiana 'Apple Blossom'

Potted plants

Many hybrid bougainvilleas make excellent pot plants, having inherited a number of desirable attributes from parent species including moderate growth rates, large showy bracts and multiple-divided cymes. These features along with recurrent, heavy flowering make the hybrids wonderful plants to grow in pots. Larger cultivars can also be grown in pots provided they are controlled by pruning.

Bougainvilleas are gross feeders, i.e., they need plenty of feeding and watering. Use an evenly balanced slow-release fertiliser (NPK 8:8:8) and apply every three months. Be careful though, too much nitrogen encourages leafy growth, to the detriment of flower production. They need a rich compost which includes some peat moss, composted animal manures and slow-release fertiliser either in the form of granules or pellets. Top up the pots with compost and fertiliser every three months.

A routine of regular pruning, feeding and watering followed by 6 weeks of flowering can result in 4 main flushes of flowers each year. If several plants are treated in this fashion but at different times, you can always have a bougainvillea in full flower. Frequent watering simulates natural, heavy rain and when followed by a dry period induces flowering.

Prune and shape your plants after they flower, keeping them to about 1 m high, and remove all the twiggy growth. Feed each pot with a balanced fertiliser that is high in potash, and contains trace elements, at the rate of one tablespoon to a 30-cm pot, and soak well with water to ensure maximum uptake of nutrients. In very hot weather, when new growth is rapid, plants may need to be watered twice each day. Watering should be gradually reduced but sufficient to prevent wilting. Foliar fertilisers can be applied at this time at about 7-10 day inter-

months. With care, bougainvilleas will grow and flower successfully in pots for over 10 years.

Repotting

Potted bougainvilleas will need repotting every second year or so, even though they can be maintained for longer periods. The procedure is to prune the top of the plant back by about one-third, turn the pot on its side and ease the root-ball out. A strong jet from the hose will help. Prune the root-ball back by about one-third and repot into rich compost. Again, this should be done in spring at the beginning of the growing season, rather than in autumn or winter.

There are many commercial potting mixes available in supermarkets and garden centres. For bougainvilleas choose a loose, friable mix, to which you should add some peat moss and fertiliser. Water-retaining elements are available and are an excellent additive. Slow-release fertilisers are ideal for potted plants. The same principles apply for whatever size pot you happen to be using. The exception is that small potted plants have yet to form a strong root system and will not appreciate the above treatment.

The old John Innes Compost recipe is ideal for bougainvilleas, using No. 1 for baby plants with new young roots and Nos. 2 & 3 for older plants. This compost originated at the John Innes Horticultural Institute in England after many years of research into plant requirements. One of the essential components is **John Innes Base Fertiliser,** which consists of:

    2 parts hoof and horn
    2 parts superphosphate
    1 part sulphate of potash.

John Innes Compost No. 1 is made from (measured by volume):

    7 parts sterilised and sieved loam
    3 parts peat moss
    2 parts coarse sand.

'Show Lady' as a potted standard.

vals. The vigorous growth of new shoots during the rainy season can be a problem because nutrient deficiency can occur, causing the leaves to turn yellow (chlorosis). This can be remedied by several applications of sequestered iron (iron chelates) and magnesium sulphate (Epsom salts).

Your plant should be in full flower about 7 weeks after pruning and will continue to flower for about 6 weeks. Once flowering starts apply a high potash fertiliser and resume heavy watering. Prune your plants again after flowering is finished, and repeat the cycle of feeding and watering. Remove any roots which appear in the drainage holes of potted plants. Remember that bougainvilleas are gross feeders, and potted specimens need topping up with rich compost every few

'Shubra' works well as a potted plant.

To each bushel (20 litres) of Compost No. 1 add 22 g garden lime and 120 g John Innes Base Fertiliser.

To make **John Innes Compost No. 2** double the quantity of Base Fertiliser for each bushel (20 litres) of Compost No. 1. With **John Innes Compost No. 3** the amount of John Innes Base Fertiliser added to each bushel is trebled.

John Innes Compost No. 1 is a standard mixture for potted plants and Nos. 2 & 3 provide progressively richer composts.

Fertilisers for potted plants

Fertilise potted plants regularly and top them up with mulch. Spray the leaves with appropriate foliar fertilisers to keep them well fed. Remember potted plants have limited soil and nutrients will be leached away when you water them.

Balanced fertiliser mix:

2 teaspoons sulphate of ammonia

3 teaspoons superphosphate

1 teaspoon potassium sulphate or potassium nitrate

½ teaspoon sulphate of iron

¼ teaspoon magnesium sulphate (Epsom salts)

This covers an area about 60 cm in diameter and gives a good boost to plant growth. Apply this fertiliser about every 3-4 weeks if you intend to use it regularly.

Standards

Bougainvilleas grown as standards are a truly magnificent sight and are well worth the time

Potted bougainvilleas used to good effect.

and effort needed to achieve the result. They can be grown in the ground or in large pots. Standards should have a single stem, but sometimes several main stems which are kept clean of shoots will help to form a more substantial umbrella of foliage. The plant should look formal and when in full flower it will become a superb ball of colour. A substantial stake is needed until the trunk is thick enough to support the crown; this could take some years.

Standard plants are sometimes available in nurseries. They are usually 3-4 years old and have a smallish crown, but if they are potted into a larger pot or planted out into the ground they should grow quickly into good-sized specimens.

If you wish to train your own standard, select a plant of good size, such as a 20 cm or larger pot with one or more strong water-shoots, or at least a long stem. Plant into either a large pot or the ground and make sure that you attend carefully to the proce-

dure outlined in the section on planting out. (See pages 50-51.)

Set either a stout stake or pipe into the ground or pot, and plant the bougainvillea beside it. A circular, triangular or square frame at the top of the stake makes an excellent support for the top growth but is not essential.

Select one or two long shoots and run it up to the height you require (about 2 m is a good height) and top it. Tie the shoot to the stake at several places; make it tight enough to hold the plant firm but not so tight that it will strangle the plant. Clip off the thorns, leaves and any lateral shoots. Now that the stem is free of thorns it will be easier to rub off any shoots which grow later.

The intention is to promote the growth of strong lateral branches which radiate out from the top of the watershoot. Keep heading back these chosen branches as they grow (nip out the tips when the shoots are 7-10 cm long) and remove spindly shoots until a solid crown of branches is achieved. New shoots can be allowed to grow and they can be wrapped around and tied back into the framework of the crown. Once the desired shape and size is achieved, the crown should be regularly pruned and spindly growth cleared out after the flowering season. Bougainvilleas are long-lived plants — up to 100 years — so it is well worth putting the effort and time into training a bougainvillea as a standard plant.

When repotting standards, the same principles apply as those used to repot any other bougainvilleas. Some people have been known to turn out the plant, beat and shake out all the soil, cut all roots back to stumps and repot successfully. As a general rule this treatment is not recommended — you can easily kill your precious plant!

Bougainvillea plants can be grown successfully in large plastic pots and are much easier to handle when repotting. However, large clay pots are aesthetically more pleasing and also help to insulate the root-ball from heat. Plastic pots absorb a lot of heat when exposed to the sun.

Potted standards should be grown in heavy concrete or clay pots to help prevent them being blown over in strong winds. All bougainvillea cultivars make suitable standard specimens because of their pendulous growth habit. In particular the cultivars 'Treasure', 'Sanderiana' and 'Vera Blakeman' look very good as formal standards.

Shrub or bush

Keep the plant cut low so that the trunk will become thick and strong to support the

'Mrs Eva' pruned as a bush.

grown shrub. Prune back long branches to maintain shape and remove watershoots which emerge from the crown. Once you have achieved the desired height and spread, prune side branches selectively to maintain size. Trim the underskirts and remove the lowest canes to avoid a hard-pruned look, but do not worry if you are forced to prune your plant hard, because new growth will compensate for the loss very quickly.

Hedge

Bougainvilleas make secure formal or informal hedges. Plants should be spaced about 1-1.5 m apart depending on the variety; smaller cultivars can be planted closer together.

To get the best results, adhere to the same general rules recommended previously for planting out. Good soil rich in humus, which

Left: 'Magnifica' as a standard.

Below: B. spectabilis as a formal hedge.

has been mixed with a generous amount of complete fertiliser in both slow-release and quick-action forms, will help the hedge grow vigorously.

Cultivars of *B. glabra*, with their compact growth habit, small leaves and long flowering period, and which respond well to clipping, make excellent formal hedges.

Great visual impact is achieved by planting the hedge with one variety only, e.g., *B. glabra* 'Sanderiana' for a purple hedge, *B. glabra* 'Alba' for a white hedge and *B. glabra* 'Easter Parade' for a lavender-pink hedge. The more upright cultivars, members of the *buttiana* group, make lovely informal hedges either in single or mixed colours. In these, when the canes grow out in the summer, they flower at the end and progressively all along the cane — a wonderful sight.

They come in a range of colours, from the original red, through orange, yellow, pink and white, and most of these now have forms with variegated leaves. *B. spectabilis* grow into larger shrubs, have seasonal flowering habits and make spectacular displays with their solid blocks of colour. This species

As hedges, bougainvilleas are both practical and beautiful.

grows well as a dense hedge and responds well to clipping.

To establish a hedge, cut away the top growth to control the height and encourage lateral spread of the plant. Each year allow a little more top growth until you have achieved the required height.

Tie the strong shoots along the length of the hedge and let their new shoots rise to fill the hedge and make a strong framework.

Developing a bougainvillea as a hedge plant

Bougainvilleas provide rapid cover for pergolas and fences.

Once the hedge is established it is important to remove all strong new canes at their source or they will spoil the shape.

After flowering is finished, light pruning will keep the hedge in shape. Thereafter remove irregular canes and branches, top and side dress annually with a slow-release fertiliser and a good thick layer (15 cm) of aged chicken litter, stable manure or compost. A further light dressing in late summer of a fertiliser high in potash helps to promote flowering.

Pergola

Plant a robust specimen of bougainvillea next to your pergola, tie a strong shoot to the end upright and remove all other lateral shoots. This main shoot will grow rapidly and extend above the pergola roof. When the shoot is about 1 m above the pergola, bend it down and tie it to the top centre framework. The lateral shoots which then grow should be direct-

ed to form a network of branches which cover the top of the pergola. Trim the shoots from time to time to keep the pergola neat and tidy, and prune after flowering; make sure that only one shoot is retained to branch out on the pergola roof. Some branches will need to be removed occasionally to reduce the density of growth on top of the pergola. The colour of the light shining through the bracts is superb, so it is advisable to keep leafy growth to a minimum.

Espalier

Plants which are grown flat, either on a wall, fence or framework, are called espaliers. Bougainvilleas are beautiful when grown in this manner, particularly against a sunny wall.

Prepare the site to be planted in the usual way and position your plant as close to the wall as possible. Attach a suitable vertical branch to the wall and let it grow as high as required. Lateral branches should be allowed to develop at appropriate spaces so that the wall may be seen between the branches, as this is part of the whole effect. Remove all unwanted shoots at their source and all forward growth, and tie the chosen lateral branches into position while they are still flexible. You can achieve any shape you desire, but the overall effect should be controlled and formal.

Clothing a steep bank

Bougainvilleas make good ground cover on banks. A fine display will be assured if the following rules are observed:

1. The bank must be in full sunshine.
2. Plant at the base of the bank, heeding all the requirements for planting out.
3. Train one or two stems up to the top of the bank and then take them horizontally in either direction. Rub off the shoots and leaves and cut thorns from the stems.

Top right: Interesting use of bougainvilleas on a high-rise building.

Bottom right: 'Partha' provides shade on a city street.

Below: Grow several different cultivars together for interesting colour effects.

A beautifully clipped standard in the Palace gardens, India.

Bougainvilleas can also be interesting subjects for bonsai.

Now the horizontal branches should put out pendulous laterals which will drape down the face of the bank to wonderful effect.

Through a tree

The bigger, rampant bougainvillea varieties like 'John Lattin', 'Meriol Fitzpatrick' and 'Scarlet O'Hara' make spectacular cascades when grown up to the top of trees.

Select a specimen with a strong stem and plant it close to, and on the sunny side of, a tree. Stop it from branching until it grows at least halfway up the tree. The shoots should now grow rapidly up through the canopy, seeking light, and once on top will produce lateral shoots which will fan out and flower in the sun. The plant must be fertilised and watered regularly because the tree will consume much of the available food and water.

Indoors

Bougainvilleas can be grown indoors provided they are given enough light. There should be enough light for the plant to cast a shadow. They can be successfully grown long term in greenhouses and in very bright sunrooms in temperate climates in various sized pots. Support will be required if you want your bougainvillea to grow up like a vine,

otherwise it can be grown as a shrubby plant. Many small potted plants are sold for seasonal indoor colour. Indeed, once the bracts have formed, they are tolerant of moderate light changes and changing temperatures, and will continue to give a good display even if you move them around.

Bonsai

Bonsai is a Japanese word which means 'planted in a bowl'. Bougainvilleas are superb subjects for the bonsai enthusiast. To grow bonsai is a fascinating art in itself and there are many good and inexpensive books on the subject. Often classes are available, if you wish to pursue this intriguing pastime.

Bonsai are generally classified according to size, attitude, number of trunks growing from a single root, the number of trees in a group planting and the kind of base on the plant. The same rules apply to bougainvilleas as to any other species being made into a bonsai. You will be absolutely delighted when your miniature bougainvillea bursts into brilliant colour. The more pendulous varieties which tend to drape naturally lend themselves to the cascade styles of bonsai but stiffer-growing varieties like 'Treasure' and 'Vera Blakeman' make very formal topiary-style bonsai.

Chapter 5

PESTS AND DISEASES

B. glabra 'Elizabeth Angus'

BOUGAINVILLEAS are favoured, both ecologically and economically, because they are relatively free of pests and diseases. You do not need to continually spray for insects and diseases to grow beautiful flowering plants. This is a desirable attribute considering the huge amounts of chemicals generally used for pest and disease control which are toxic and pollute the environment.

Bougainvilleas grown in full sun, which is their preferred situation, will be much more resistant to attack from insects and pathogens, whereas plants growing in the shade will be more vulnerable to attack.

Leaf spot

In prolonged wet conditions in the tropics, leaf spot can be a problem, even for bougainvilleas, and when the conditions are very severe, leaf spot can account for 80 percent of leaf fall in young plants.

The major cause of leaf spot worldwide is the bacteria *Pseudomonas stizilobii*, which is often accompanied by a fungal leaf spot which likes the same conditions. The fungi that cause disease are *Cercosporidium bougainvillae*, *Colletotrichum dematum bougainvillae* and *Gladosporum arthrinoides*.

The appearance of small, reddish, rounded spots on the leaves is a characteristic early symptom of infection by *Pseudomonas stizilobii*; the spot is usually surrounded by a 'halo' of pale green which merges into the normal leaf colour. The spots rapidly expand to become irregular dark patches which are bounded by small veins, and when the attack is severe, the patches become almost black and distort the young leaves, this leads to severe defoliation. As the dry season approaches, the new growth can still become infected but usually less severely. Leaves which remain on the plant until the following year are the source of infection for when conditions next become suitable.

Leaf spot can be a problem in wet, humid areas.

The disease is transmitted in water and can be prevented by removal of all dead leaves and by reducing crowding between plants. Plants grown in greenhouses watered by overlapping sprinklers are particularly susceptible to the spread of this disease. Cool, showery nights and late afternoon watering in cool weather often result in leaves which remain wet overnight and can also increase the susceptibility of your plants to the disease. Spray with a fungicide at the first sign of the disease. These can be effective for long periods and will control a wide range of diseases, in addition to fungi. They are sold under various trade names, but the active ingredients include benomyl, captan, maneb, mancozeb and zineb. It is essential to follow the instructions on the packet exactly and use the recommended concentrations.

Mites

Marginal leaf curl, where the leaves cup and distort, is caused by mites. The condition is not usually severe and by the time the distortion is apparent the mites have been and gone. It is thought the mites secrete a hormone in their saliva which interferes with plant growth. The mites inject their saliva when they feed on the young tips of the plants. Mite infestations usually occur during warm, humid weather and new growth which appears at other times will be normal.

Spray severe cases regularly at intervals of 10-14 days with a systemic acaricide and thereafter at the onset of warm weather. This will prevent further infestation. If the disease is not severe, either ignore it or cut away the affected branches.

The distortion of this leaf has been caused by mite attack.

Aphids

Aphids sometimes infest new growth and flowering shoots, causing distorted growth and sometimes even the death of shoot tips. Infestation can become particularly bad in humid glasshouse conditions. Spray with a systemic acaricide which also controls mites, scale and grasshoppers. Maldison is an effective spray with a short residual action and is not toxic to animals and humans. Again, adhere to all instructions on the label.

Other insects

There are some insects which can be a nuisance in the U.S.A., and these are controlled by recommended insecticides. One of these insects is the soft scale *Coccus hesperidium*, which attacks plants grown in both the greenhouse and open garden. In Florida, the caterpillar *Asciordes gordinalus* and the beetle *Amphericus cornatus* Pallas can also be problems. The beetle is commonly known as the bougainvillea beetle.

In the past, ants were controlled by toxic chemicals with long residual action. These have now been removed from the market. Check with your local garden centre about the current methods to control ants. Dust insecticides are useful when repotting ant-ridden specimens; dust the new soil, the old root-ball and the top of the pot. Ants are a nuisance in pots as they make passages in the soil and also make the soil water-repellent.

Earthworms in pots

Earthworms, those magical little creatures that we love to have in our soil, are not so welcome when in your potted plants. They are a nuisance rather than a pest because they convert potting mix into a heavy sludge; rotenone (Derris Dust) at the rate of 28 g of 1 percent dust to 1 sq. m of soil will evict them.

Chapter 6

PROPAGATION

Variegated cultivar 'Raspberry Ice'

IT is most rewarding to propagate your own plants. Most bougainvilleas 'strike' quite easily and those that are difficult to grow by cutting can either be air-layered (marcottage) or layered. Bougainvilleas can also be grafted and budded.

Hardwood cuttings

These are pieces of woody stem about 15-20 cm long and 10-15 mm in diameter taken in late winter. They are cut on an angle just beneath the node at the basal end of the cutting, and just above the node at the top end. The cutting is either dipped or dusted with root-promoting hormones and planted to about two-thirds of its length into a 50:50 mixture of sand and peat moss in a bed in the open. Root-promoting hormones are not essential, but they will increase the percentage of cuttings that strike. They also promote a better root structure and therefore a stronger plant. Take many more cuttings than

you need because only about one in three will grow into a good plant. It can take up to 3 months before you get really good root and leaf growth, but you will have a strong and sturdy plant.

Semi-ripe cuttings

These cuttings are taken in spring from the previous season's growth, when the new buds are showing signs of growth. The cuttings should be 15-20 cm long, as thick as a pencil, and treated the same as hardwood cuttings. Eight to 10 cuttings are placed in a 12 cm pot filled with a 50:50 coarse sand and peat moss mix and kept either in the greenhouse or warm bush-house. The cuttings are ready to separate and pot when there is strong aerial growth and thick white roots show through the drainage holes. Up-end and remove the plants from the pot carefully. Gently break open the soil mass, taking care not to damage the roots. Small bougainvillea

Hardwood cutting material.

Semi-ripe cutting material.

plants have very delicate and brittle roots, which need careful handling. Separate the cuttings and plant them into individual 12 cm pots. Place in a warm, sheltered position until the baby plant has recovered and is growing well. When the plant has become well developed, pot up into a larger container. At the beginning of winter it is safer to leave this further repotting until the spring; in any case do not overpot small plants.

Layering

Many varieties which do not root easily from cuttings can be successfully layered. Layering is a simple method for propagating small numbers of plants, which does not need either special skills or equipment. Layering allows the propagation of larger plants in a shorter time than it takes to grow them from cuttings.

Root formation during layering is stimulated when the flow of nutrients and auxins from the shoot tips to the roots is interrupted;

the natural stem flow of auxins is from the apex to the base. Auxins and nutrients build up where the stem is cut and induce roots to form there. Root formation is facilitated by darkness, so the point of layering should be either buried or mounded with soil. Soil which retains moisture and is full of humus will help roots to develop, particularly when the cut stem has also been treated with root-promoting hormones.

The stem has to be wounded because the injury causes the hormones to accumulate, which encourages roots to form. Make a slit about 35 mm long through the centre of the stem just below a node or leaf junction and insert a small foreign body (e.g., a matchstick) to keep the cut edges apart. Cover the layer with soil, then plastic weighed down with stones to stop soil from blowing and washing away.

Once the layered stem starts to grow, cut it free from the parent plant and pot it up carefully, taking care not to damage the roots.

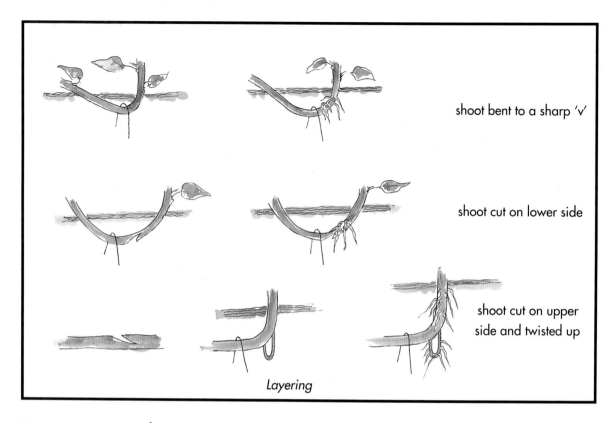

shoot bent to a sharp 'v'

shoot cut on lower side

shoot cut on upper side and twisted up

Layering

Marcottage or air-layering

Marcottage or air-layering is a good method to propagate plants, particularly if only a few plants are required. It also means that a larger plant can be acquired immediately. It was commonly used as a propagation method in the past, before modern techniques took hold. It is useful for hard to propagate plants.

Select a pencil-thick stem and cut it on an angle just below a node, three-quarters through its thickness. Dust the cut with root-promoting hormone in powder form and keep the edges open with a piece of gravel, coarse sand or other suitable material about 2 mm in diameter. Wrap a generous amount of thoroughly moistened sphagnum or peat moss around the cut, cover with plastic film, and tie the centre and each end to keep the moss in firm contact with the cut surface. The layer is successful when roots show in the plastic cocoon and new growth starts to appear at the top; this usually takes about 1 month. If the moss appears to be drying out, untie the top and pour in a little water. The layered branch will require some support because the strength of the cut stem is much weakened. Pot up the new plant in the usual fashion.

Grafting

Grafting is the art of joining areas (the cambium layer) of a cutting to an understock to grow as one plant. The cambium layer is the tissue which lies between the bark and the central wood of a plant. Grafting is especially useful when you want to grow delicate varieties which have fragile root systems. It is also used when gardeners want multiple varieties on one plant. Grafting plants can be traced back to ancient times. Mixtures of wet clay and dung were used in those days to cover graft unions.

A successful graft aims to fit two pieces of living plant tissue together so that they will unite and grow as one plant. The short piece of stem material which is to form the leafy

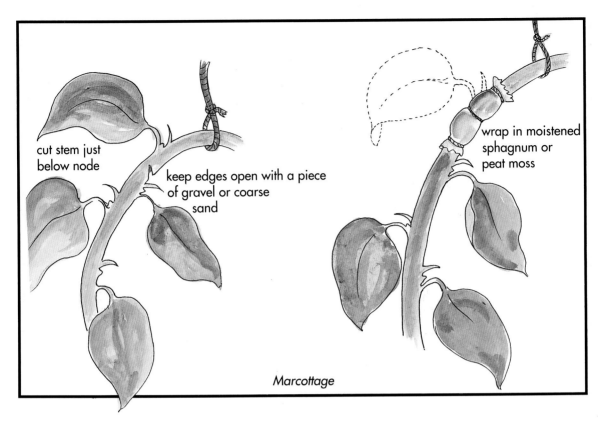

cut stem just below node

keep edges open with a piece of gravel or coarse sand

wrap in moistened sphagnum or peat moss

Marcottage

canopy of the plant is called the scion. It should be of the desired cultivar and free of disease. The stock or rootstock is the lower part of the graft and could be a seedling or a rooted cutting of an existing established plant. Usually a desirable cultivar which is difficult to propagate because it produces a poor root system of its own is grafted onto a stock which grows a vigorous root system. To graft successfully, the scion and rootstock must be compatible; the cambiums of the stock and scion must be in close contact with each other, and the cut surfaces held together tightly by wrapping. The wound needs to heal rapidly if the scion is to be supplied with nutrients and water by the rootstock.

All cut surfaces should be covered with grafting wax once the join is made, to prevent moisture loss. Care of grafted plants must be maintained after grafting; support the scion, remove shoots from the rootstock, and put the grafted plant in a humid environment so that the scion does not dry out.

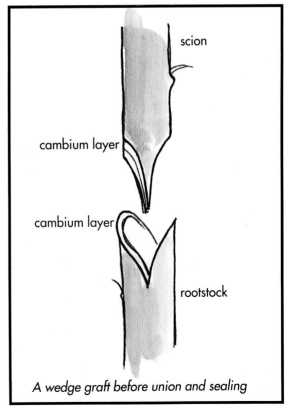

scion

cambium layer

cambium layer

rootstock

A wedge graft before union and sealing

The spectacular results of two cultivars grafted onto one root stock.

Some variegated bougainvillea cultivars have little or no chlorophyll in their leaves and are extremely difficult to grow from cuttings and therefore need to be grafted onto a vigorous rootstock to make new plants.

Budding

Another aspect of growing bougainvilleas as a hobby is the growth of several varieties on a single plant to make exciting and colourful displays. This can be done by grafting and also by the technique of budding. This involves the insertion of a bud (the scion) under the edges of a T-shaped cut in the bark of the plant (the rootstock) to which you wish to add the new variety. A clean, sharp knife should be used and it is a good idea to have a few practise runs.

Make a 2 cm T-shaped cut in the bark of the recipient plant and insert the donor bud

cambium exposed and prepared for union

bound and ready for sealing with wax

Approach graft

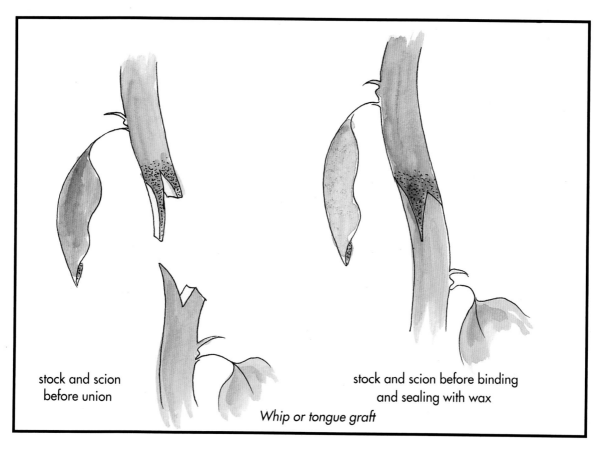

stock and scion
before union

stock and scion before binding
and sealing with wax

Whip or tongue graft

under the flaps. Tie with either raffia or graft-ing tape so that there is no movement. Seal the area with grafting wax or a similar com-pound to exclude air. If the bud is still green and healthy looking after 3 weeks it will prob-ably be a successful union.

The bud should be selected from the cur-rent year's wood and it is important the piece chosen does not dry out. Should you take a large cutting with a number of eyes in it, cover it with a wet cloth while you make the incision in the rootstock. Leave about 1 cm above and below the eye when you cut it from the stem, and remove thorns and pithy pieces of wood from behind the eye by gen-tly squeezing the bud piece. Slip the bud, right way up, in behind the flaps in the T-cut and bind securely to prevent movement. Seal with grafting wax. If your first efforts are not successful, keep trying until you have mas-tered the technique.

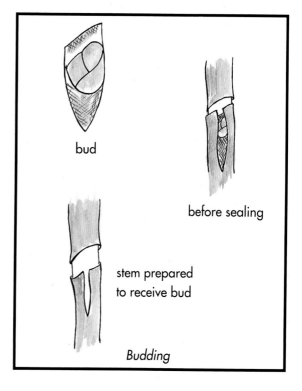

bud

before sealing

stem prepared
to receive bud

Budding

Chapter 7

GENETICS AND BREEDING

B. spectabilis 'Rubra Variegata'

IN all living creatures it is usual to have a double set of chromosomes, all in pairs. These chromosomes act as a genetic 'code' determining the characteristics of the individual. The individual inherits one set of chromosomes from each of its parents, and so also inherits some of the characteristics of each parent. In biology, the usual double set of chromosomes is referred to as the diploid state (2n). (Haploidy means an individual has only one set of chromosomes (n), triploidy means there are three sets (3n) and tetraploidy four sets (4n). Various other forms of multiple sets of genes are also seen.) While not necessary to grow bougainvilleas, or indeed any plant, a basic knowledge of genetics helps to explain why certain plants inherit certain characteristics and makes the process of breeding new varieties more interesting.

Fertility and interbreeding

Bougainvilleas have n = 17 chromosomes and the majority of cultivated varieties are diploid, (2n=34). Some, however, are triploid (3n=51) and there are natural as well as induced tetraploids (4n=68).

In self-pollination, pollen is transferred from the stamens of a flower to the stigma of the same flower. Cross-pollination means transfer of pollen from the stamens of one plant to the stigma of another flower. In general, pollination is successful only between plants of the same species, but hybrids can result when pollen successfully fertilises plants of an allied species. Self-pollination is prevented in many flowers by devices such as discharge of pollen before the stigma of the same flower is receptive. Such mechanisms promote cross-pollination and increase variation within the species. Genetic variation

B. spectabilis 'Splendens'

enhances a species' ability to adapt to different and changing environments.

Flowers, which are specialised reproductive organs, are central to the survival and evolution of the flowering plants. There is some specialisation into male and female parts to different degrees, depending on the species.

Like most flowering plants, *Bougainvillea* have a brightly coloured structure (the flower bracts are actually a set of specialised leaves rather than petals of a flower) to attract insects that serve to transfer pollen from one flower to the stigma of another, as described below.

The genetic information that dictates the nature of the plant and its growth, development, reproduction and the business of survival from day to day is carried in the form of genes, which act as little 'messengers' grouped together in long strands called chromosomes. The form and number of each chromosome are identical in organisms with the same heredity, and within different cells of the same individual (with rare exceptions). Reproduction means copying the genetic material and packaging it for exchange with another organism that may have slightly different genes on otherwise identical (and therefore compatible) chromosomes.

Interbreeding has resulted in a range of largely infertile *Bougainvillea* cultivars. Fertility has been restored to some through treatment with mutagenic compounds which has produced tetraploids (plants with two pairs of chromosomes). This has opened up the possibilities of further breeding of new cultivars. Genetic manipulation is the subject of ongoing research, especially in India, and opens up new possibilities such as smaller-growing varieties — perhaps even a thornless variety may be possible.

B. glabra 'Cypheri'

The issue of fertility is clouded by a lack of comprehensive knowledge, inconsistent nomenclature and reports on different cultivars, and the occurrence of variation in response to climatic and environmental change. For example, it has been reported in India that *B. x buttiana* and all the bud-sports examined on this hybrid have sterile pollen and seed (Sharma & Battachaya 1960, Zadoo et al 1974-82), but seedlings from *B. x buttiana* 'Mrs Butt' have been reported from Kenya and South Africa.

It is well known that seed-set and fertility are largely influenced by climate (usually favoured by warm, dry conditions) and it is possible that a planned breeding programme may only achieve what has occurred naturally in the past.

Inter-species variation

It is generally held that bougainvilleas are self-sterile and that most modern varieties, certainly those in widespread commercial horticulture, are completely sterile. The occurrence of new varieties are limited and they now come chiefly from bud-sports (a chance mutation, see page 82). The occurrence of bud-sports has increased with treatment of plants by irradiation. Even a fertile specimen produces a small amount of seed from thousands of flowers. Seedlings which I have grown from naturally fertilised *B. spectabilis* have developed into plants with a wide variety of colours and bract sizes, and with differing degrees of furriness of the foliage. There will be much unknown variation among plants grown from seed taken from different places. It is noteworthy that seeds brought into various countries and grown on have resulted in plants which show only subtle variations in bract colour and form, having the basic characteristics of exist-

Opposite: B. spectoperuviana 'Shubra'

B. x buttiana 'Mrs Butt'

ing varieties. Unfortunately, many of these seedlings are given new names and are claimed as unique varieties without adequate attention to the true distinguishing features of leaf, flower, thorn type, habit and bract colour.

Seedlings of *B. glabra* show a much wider range of differences in size and shape of leaf and bract than do the seedlings of *B. spectabilis*, but the colour is always within the range of purples and mauves. *B.* 'Alba', however has seedlings which are very similar in appearance and all have white bracts of variable size.

B. spectabilis and *B. glabra* are similar in essential form and structure; the major difference is that *B. spectabilis* is hairy. Both species show considerable variability. *B. peruviana* differs from *B. glabra* and *B. spectabilis* in leaf shape, division of cyme, shape of flower tube, and there is no variability within the species; indeed there is only one representative cultivar, called *B. peruviana* 'Lady Hudson' (syn. 'Princess Margaret Rose').

Natural mutations occur spontaneously

Opposite: B. spectoperuviana 'Mrs H.C. Buck'

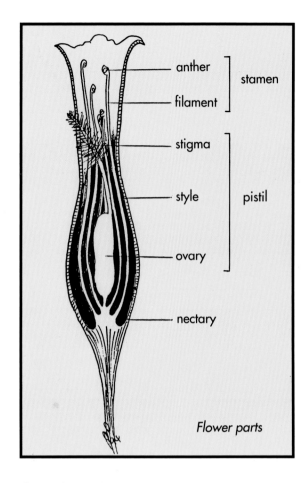

Flower parts

those which include *B. peruviana* as one of the parent species. Well-known examples include *B.* x *buttiana* 'Mrs Butt' (*glabra* x *peruviana*) which has produced sports of a number of colours and leaf variegations, as well as plants with incomplete flower parts.

'Mrs H.C. Buck' (*spectabilis* x *peruviana*) has produced a bicoloured mutation, as well as white and variegated leaf forms.

Numerous other sports are known, but some have uncertain parentage. It is interesting that apparent mutations sometimes vanish as mysteriously as they appear, and that colour changes are usually lighter (e.g., red to orange to gold to pink and white) and may even be seen serially in successive sports. Occasionally, a sport back to the original colour may be seen.

The flower

Bougainvillea flowers are always in groups of three and each one is attached to the upper surface of a brightly coloured bract. The bracts serve to attract pollinators and act as a wing in seed dispersal. One flower in the group of three opens first, and this flower is considered the terminal one.

The flower structure is in the form of a single tube, with a slight constriction in the middle, above a swollen base. The funnel-shaped upper part of the flower opens out into a five-rayed 'star'. There are five ridges along the length of the flower tube. The five points are considered to represent the sepals; there are no petals, only a thin, wrinkled, white or cream membrane which joins the five points of the star to form the conspicuous part of the flower. A few stamens can be seen in the opening of the flower tube. There are about eight stamens of variable length and their filaments spring from a little cup (the nectary) at the base of the ovary. The ovary is not symmetrical at the top and carries a tapering brush-like stigma. The stigma is surrounded by the filaments of the stamens with the anthers all above the stigma. There is a single

throughout the world; wherever large numbers of plants are being produced, bud-sports will occur. This has led to multiple names for the same cultivar and has contributed to the confusion over the names of bougainvillea cultivars.

Bud-sports

A spontaneous and unpredictable genetic change in an organism gives rise to a mutation which is generally permanent and inheritable. In the plant world, changes in the basic genetic information (a true mutation) or changes (sometimes temporary) in the way that information is expressed may lead to differences in leaf or flower — these are called sports or bud-sports.

Many of the cultivars known in the genus *Bougainvillea* come from sports, particularly from the more complex hybrids, and usually

B. spectabilis ' Wallflower'

ovule at the base inside the flower and it is this which develops into a seed, which when ripe looks not unlike a grain of wheat. Seed set is determined by the presence of pollinators and certain climatic conditions. Some cultivars are fertile and many are not. It takes about 30 days from pollination for the fruit to ripen.

Manual pollination

The pollination of *Bougainvillea* flowers is not difficult. To do so, take a flower bud that is almost ready to open; a cut with a razor blade about halfway down the upper part of the flower tube will leave the stigma intact while removing some of the anthers; remove the remaining anthers with forceps but leave the stigma untouched. Cover the flower overnight, dust pollen from the desired parent onto the stigma and re-cover the flower to prevent casual pollination.

Pollinate selected flowers early in the morning when the weather is dry. Label the potential seed flower with information about the experiment.

Seeds become ripe in about 30 days and thereafter germinate quickly. The seed is viable for some time.

INDEX OF CULTIVARS

Small-growing cultivar 'Rosenka'

Double-bracted cultivar 'Cherry Blossoms'

B. spectoperuviana 'Golden Summers'

Left: Inter-specific hybrid 'Killie Campbell'
Below: Small-growing cultivar 'Mrs Eva'

Variegated cultivar 'Ratana Pink'

Left: 'Roseville's Delight'
Below: Variegated cultivar 'Thimma'

GLOSSARY

Accuminate tapering to a point

Acute sharply pointed but not drawn out

Adnate joined to

Adventitious in other than the usual place

Air-layering *see* Marcottage

Angiosperma the group of plants which possess flowers

Anther pollen-bearing top of stamen

Apex tip of organ (e.g., leaf)

Apical relating to the apex

Apiculate with a short, stiff point

Arborescent becoming tree-like, woody

Asexual propagates without sex, so no new genetic material is introduced

Attenuate becoming narrow, tapered

Aurea golden

Auxin plant hormone which regulates growth. Used for stimulating formation of adventitious roots

Axil the point just above the leaf where it arises from the stem

Bagasse residue after extracting juice from sugarcane or beet

Basal at base of organ

Betacyanins red-purple pigment found in plants

Betaxanthins yellow-orange pigments found in plants

Blade the expanded, flattened part of a leaf

Bract modified leaves intermediate between flower and normal leaves, often coloured

Bud-sport a spontaneous genetic change, permanent or temporary; a mutation

Calyx outer circle or cup of floral parts (usually green)

Cambium cylindrical layer of actively dividing cells which give rise to secondary (woody) tissues, causing increase in girth of stem

Chimera organism with two or more genetically distinct tissues

Chlorosis yellowing of leaf

Chlorophyll complex green pigment found in green plants, active in the process of photosynthesis

Chromosomes microscopic rod-like bodies in the plant cell, made up of genes and bearing the 'blueprint' for reproduction

Ciliate fringed with eyelash-like hairs

Clasping with the leaf or bract surrounding the stem

Compound similar parts aggregated into a common whole

Concave curved inwards; hollowed out

Connate united in one body

Convex curved outwards

Cordate heart-shaped

Coriaceous leathery

Corolla complete circle of petals

Crenate with teeth rounded, scalloped

Cultivar special genetic form originating in cultivation

Cuneate wedge-shaped, triangular

Cuspidate tipped with a short, stiff point

Cyme type of inflorescence in which the flower stem elongates from a series of lateral branches

Defoliate lose leaves/foliage

Decurrent leaf blade extends down petiole

Deltoid triangular

Diploid having two sets of chromosomes

Divided separated at the base

Doubling increased number of petals or bracts beyond the norm

Elliptical oblong with the widest point at centre

Elongate drawn out in length

Emasculate remove masculine flower

Entire margin without toothing or division

F1 hybrid first-generation hybrid obtained by artificial cross-pollination between two genetically dissimilar parents

F2 generation second generation of a given cross. Usually obtained through self-pollination of the F1 hybrids so that there is genetic variation among individual plants

Floriferous bearing flowers

Friable crumbly texture

Fusiform spindle-shaped, rounded and tapering from the middle towards each end

Glabrous smooth, not hairy or rough

Haploid having one set of chromosomes

Hirsute hairy, with long rather stiff hairs

Hybrid plants resulting from a cross between parents that are genetically different

Inflorescence the flowering part of a plant

Internode space between two nodes

Laciniate slashed into narrow, irregularly-pointed lobes

Lanceolate lance-shaped, tapering towards tip

Lenticel raised corky projections on young bark giving vent to breathing pores

Limb the border or expanded part of the corolla above the throat

Linear narrow and flat, margins parallel

Lobe any projection of a leaf, rounded or pointed

Marginal at the edge

Marcottage a propagation technique

Meristem region of plant tissue where cells are actively dividing, e.g., cambium layer, shoot or root tips and at nodes

Morphology the form and structure of plants

Mutant form derived from sudden genetic change, which continues in subsequent generations

Node a joint in a stalk where the leaves arise

Oblanceolate broad end near tip, tapering toward base

Ovate a leaf broadest near base, tapering upward

Panicle an open and branched flower cluster

Pedicel stalk of each flower and cluster

Peduncle primary flower stalk

Pendant hanging from its support

Perianth the calyx and corolla

Petiolate furnished with a petiole

Petiolule branch of a petiole

Pistil female organ of a flower, consisting of ovary, style and stigma

Polymorphic variable as to habit and form

Puberulent minutely pubescent

Pubescent downy

Raceme elongated, simple inflorescence with stalked flowers

Recurved bent backward or downward

Rootstock the plant to which the scion is grafted

Rosea rose coloured

Rugose covered with wrinkles

Scandent climbing

Scion the shoot or piece of stem material to be grafted onto the rootstock

Sepal each segment of the calyx

Simple leaf with one blade

Single flower flower with one set of petals

Spectabilis spectacular, remarkably showy

Spermatophyte seed-bearing plant

Stamen pollen-bearing male organ

Stellate star-like in form

Stigma that part of the pistil or style which receives the pollen

Style connecting stalk between ovary and stigma

Sub-cordate indented slightly

Subtend to extend under or be opposite to

Synonym (syn.) a word having the same meaning as another

Taxonomy the name and classification of plants

Tetraploid having four sets of chromosomes

Throat opening of the flower tube

Tomentose densely covered with matted wool

Triploid having three sets of chromosomes

Truncate as if cut off at the end

Tube united portion of calyx or corolla

Undulate wavy margined

Variety a type of modification, subordinate to species

Vegetative growing in some way other than from seed or spore, e.g., cuttings, layers

Versicolour variously coloured, changing colour

Watershoot strong, succulent shoot which arises from latent buds on woody branches

Xanthophyll yellow pigment in plants

Zygote fertilised egg

Approximate metric conversions

To Convert	to	Multiply by
metres (m)	feet	3.28
centimetres (cm)	inches	0.40
millimetres (mm)	inches	0.04
cubic metres (m^3)	cubic yards	1.31
litres (l)	gallons (US)	0.26
litres (l)	gallons (Imp)	0.22
grams (g)	ounces	0.24
kilograms (kg)	pounds	2.20

BIBLIOGRAPHY

Bailey, L.H., "*Bougainvillea*", In: *Manual of Cultivated Plants* (14th ed.), 1975, pp. 357-8.

Bannochie, I., "*Bougainvillea* as Pot Plants", In: *The Garden*, Vol. 104, pp. 11-14.

Bhat, R.N., New *Bougainvillea* Cultivars, In: *Indian Horticulture*, Jan. 1986, pp. 27-8.

Bor, N.L. and Raizada, M.M., *Nyctaginaceae*, In: *Some Beautiful Indian Climbers and Shrubs*, 1954, pp. 267-74.

Campbell, Killie, "The Waywardness of Bougainvilleas", In: *Gardeners in Conference*, SABC, 1956.

Delap, H.A., *Bougainvillea News*, Vol. 1, 1950; Vol. 2, 1951; Vol. 3, 1952-3.

Everett, T.H., "Bougainvilleas", In: *The New York Botanical Garden III, Encyclopedia of Horticulture*, Vol. 2, pp. 472-4.

Gillis, W.T., "Bougainvilleas of Cultivation", In: *Baileya*, Vol. 20, pp. 34-41.

Herklots, G., "*Nyctaginaceae*", In: *Flowering Tropical Climbers*, 1976, pp. 133-5.

Holttum, R.E., "The Cultivated Bougainvilleas I: Horticulture", In: *Malay Agri-Hort Magazine*, Vol. 12, pp. 2-10.

——————. "The Cultivated Bougainvilleas II: *B.* x *buttiana* and its cultivars and hybrids", In: *Malay Agri-Hort Magazine*, Vol. 12 (3), pp. 2-11.

——————. "The Cultivated Bougainvilleas III: *B. glabra*", In: *Malay Agri-Hort Magazine*. Vol. 13 (1), pp. 25-36.

——————. "The Cultivated Bougainvilleas IV: *B. spectabilis* and its varieties', In: *Malay Agri-Hort Magazine*, Vol 13 (1), pp. 13-22.

——————. "The Cultivated Bougainvilleas V: The Recorded Hybrids", In: *Malay Agri-Hort Magazine*, Vol. 13 (2), pp. 66-72.

——————. "*Bougainvillea*", In: *A Practical and Scientific Encyclopaedia of Horticulture*", 1969, pp. 204-5.

——————. "*Bougainvillea*", In: *Edwin Menninger's Flowering Vines of the World; An Encyclopaedia of Climbing Plants*", 1970.

Lim, G. and Wang, P.K., "*Cercosporidium bougainvilleae* on *Bougainvillea* in Singapore", In: *Korean Journal of Plant Protection*", Vol. 1 (2), pp. 121-3.

McDaniels, L.H., "A Study of *Bougainvillea* (*Nyctaginaceae*)", In: *Baileya*, Vol. 21 (2), pp. 77-100.

Ohri, D. and Zadoo, S.N., "Cytogenetics of Cultivated Bougainvilleas: X Nuclear DNA content", In: *Z. Pflanzenzuchtg*, Vol. 88, pp. 168-73.

——————., "Cytogenetics of Cultivated Bougainvilleas 1X. Precocious Centromere Division and Origin of Polyploid Taxa", In: *Plant Breeding*, Vol. 97, pp. 227-31.

——————., "Cytogenetics of Cultivated Bougainvilleas VIII. Cross Compatability of Relationships and Origin of *Bougainvillea* 'H.C. Buck' Family", In: *Z. Pflanzenzuchtg*", 1978, pp. 182-6.

Pal, B.P., Zadoo, S.N. and Khoshoo, T.N., "Genesis of *Bougainvillea* Shubra", In: *Indian Journal of Horticulture*, Vol. 33(1), p. 60.

Pancho, J.V. and Bardenas, E.A., 'Bougainvilleas in the Philippines", In: *Baileya*, Vol. 7, pp. 91-101.

Pancho, J.V., "Notes on Double Bracted Bougainvillea in the Philippines", In: *The Lal-Baugh*, Vol. VII(2), pp. 30-2.

Preston, F.G., *Royal Horticultural Society of Gardening*, Vol. 1, 1951, p. 302.

Watson, D. and Criley, R.A., *Bougainvilleas*, University of Hawaii, 1973.

Webber, H.J., "*Bougainvillea*", In: *The Standard Encyclopaedia of Horticulture*, Vol. 1, pp. 533-4.

Wood, P., "*Bougainvillea*", In: *A Comprehensive Guide to Gardening in Rhodesia (Zimbabwe)*, Vol. 1, pp. 108-14.

INDEX